Man-made Wonders

EXPLORE AMERICA

Man-made Wonders

THE READER'S DIGEST ASSOCIATION, INC.
Pleasantville, New York / Montreal

MAN-MADE WONDERS was created and produced by ST. REMY MULTIMEDIA INC.

STAFF FOR MAN-MADE WONDERS
Series Editor: Elizabeth Cameron
Art Director: Solange Laberge
Editor: E. W. Lewis
Assistant Editor: Neale McDevitt
Photo Researcher: Linda Castle
Cartography: Hélène Dion, David Widgington
Designer: Anne-Marie Lemay
Research Editor: Robert B. Ronald
Copy Editor: Joan Page McKenna
Contributing Researchers: Olga Dzatko, Suzanne Léveillé
Index: Linda Cardella Cournoyer
Production Coordinator: Dominique Gagné
Systems Director: Edward Renaud
Technical Support: Jean Sirois
Scanner Operators: Martin Francoeur, Sara Grynspan

ST. REMY STAFF
PRESIDENT, CHIEF EXECUTIVE OFFICER: Fernand Lecoq
PRESIDENT, CHIEF OPERATING OFFICER: Pierre Léveillé
VICE PRESIDENT, FINANCE: Natalie Watanabe
MANAGING EDITOR: Carolyn Jackson
MANAGING ART DIRECTOR: Diane Denoncourt
PRODUCTION MANAGER: Michelle Turbide

Writers: Elizabeth Cameron—New York's Skyscrapers
Rod Gragg—Gateway Arch
Jim Henderson—Louisiana Superdome
Pierre Home-Douglas—Kennedy Space Center
Rose Houk—Glenwood Canyon Drive
K. M. Kostyal—C&O Canal
Steven Krolak—Golden Gate Bridge, Hoover Dam,
Monterey Bay Aquarium
Alfred LeMaitre—Brooklyn Bridge

Contributing Writers: J. R. Adams, Adriana Barton,
Alfred LeMaitre

Address any comments about *Man-made Wonders*
to Editor, U.S. General Books, c/o Customer Service,
Reader's Digest, Pleasantville, NY 10570

READER'S DIGEST STAFF
Group Editorial Director, Travel: Linda Ball
Senior Editor: Fred DuBose
Editors: Tom Ranieri, Alexis Lipsitz
Art Editor: Martha Grossman
Production Supervisor: Mike Gallo

READER'S DIGEST GENERAL BOOKS
**Editor-in-Chief, Books and Home
Entertainment:** Barbara J. Morgan
Editor, U.S. General Books: David Palmer
Executive Editor: Gayla Visalli
Managing Editor: Christopher Cavanaugh

Opening photographs
Cover: Empire State Building, New York
Back Cover: Tehachapi Wind Farms, California
Page 2: Hoover Dam, Arizona/Nevada
Page 5: Red Rocks Amphitheatre, Colorado

Library of Congress Cataloging in Publication Data

Man-made wonders.
        p.   cm.—(Explore America)
      Includes index.
      ISBN 0-7621-0052-4
      1. Civil engineering—United States.   2. Structural engineering
      —United States.   3. Architecture—United States.   4. Building—
      United States.   I. Series.
      TA148.M36   1998
      917.304929—dc21   97-48425

CONTENTS

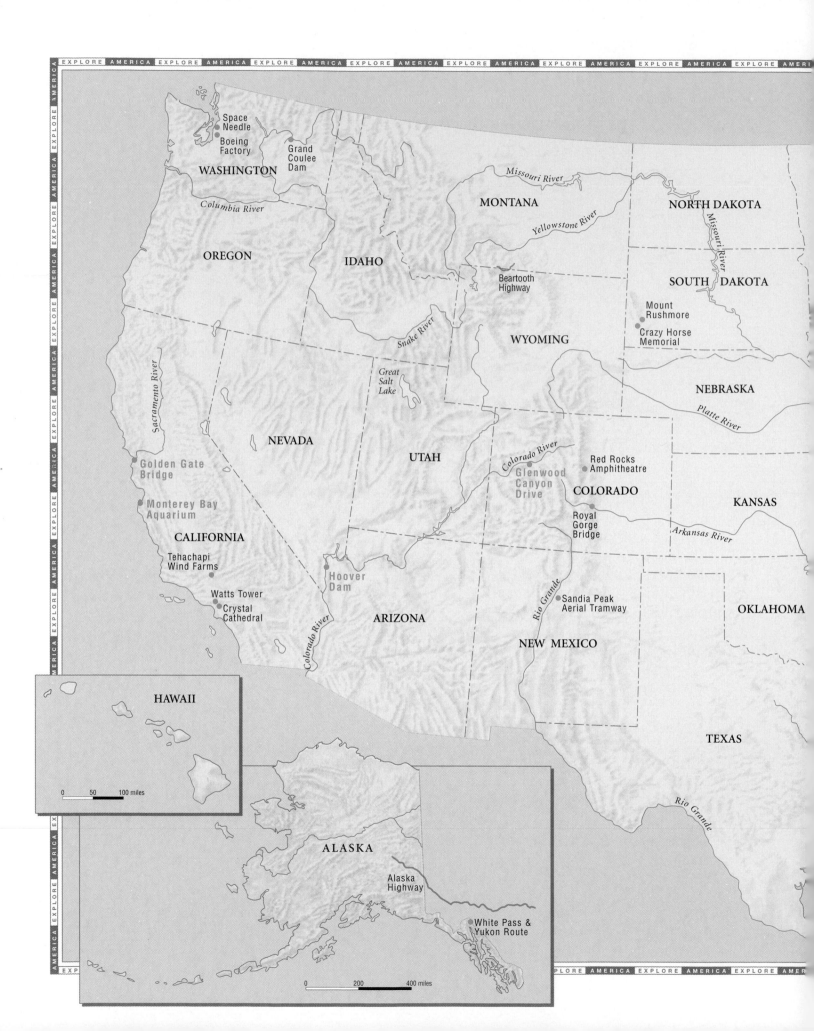

Space
Needle
Boeing
Factory
Grand
Coulee
Dam

WASHINGTON

Columbia River

Missouri River

MONTANA

NORTH DAKOTA

Missouri River

Yellowstone River

OREGON

IDAHO

Beartooth
Highway

SOUTH DAKOTA

Snake River

WYOMING

Mount
Rushmore

Crazy Horse
Memorial

Sacramento River

Great
Salt
Lake

NEBRASKA

Platte River

NEVADA

UTAH

Colorado River

Red Rocks
Amphitheatre

Golden Gate
Bridge

Glenwood
Canyon
Drive

COLORADO

KANSAS

Monterey Bay
Aquarium

Royal
Gorge
Bridge

Arkansas River

CALIFORNIA

Tehachapi
Wind Farms

Hoover
Dam

Watts Tower
Crystal
Cathedral

ARIZONA

Rio Grande

Sandia Peak
Aerial Tramway

OKLAHOMA

Colorado River

NEW MEXICO

HAWAII

TEXAS

0   50   100 miles

Rio Grande

ALASKA

Alaska
Highway

White Pass &
Yukon Route

0   200   400 miles

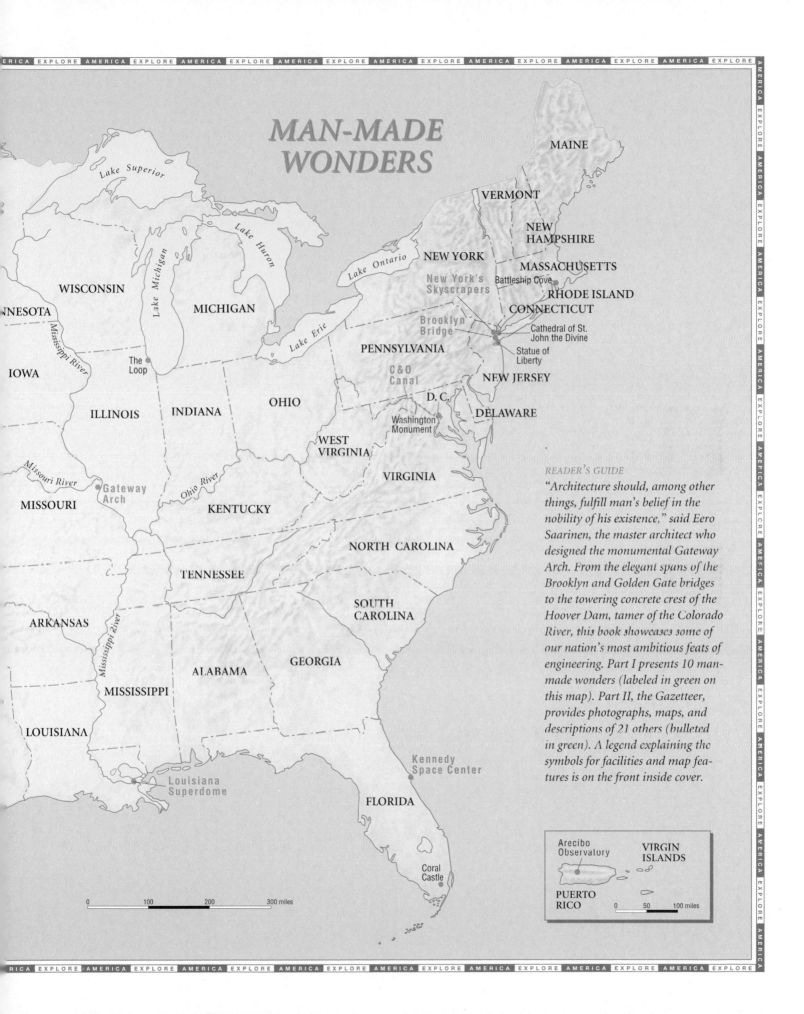

MAN-MADE WONDERS

MAINE

Lake Superior

VERMONT

NEW
HAMPSHIRE

NEW YORK

Lake Michigan

MICHIGAN

Lake Huron

Lake Ontario

WISCONSIN

MASSACHUSETTS

Battleship Cove

NNESOTA

RHODE ISLAND

New York's
Skyscrapers

CONNECTICUT

Lake Erie

Brooklyn
Bridge

Cathedral of St.
John the Divine

Mississippi River

The
Loop

PENNSYLVANIA

Statue of
Liberty

IOWA

C&O
Canal

NEW JERSEY

ILLINOIS

INDIANA

OHIO

D.C.

DELAWARE

Washington
Monument

WEST
VIRGINIA

Missouri River

Gateway
Arch

Ohio River

VIRGINIA

MISSOURI

KENTUCKY

NORTH CAROLINA

Mississippi River

TENNESSEE

ARKANSAS

SOUTH
CAROLINA

GEORGIA

ALABAMA

MISSISSIPPI

LOUISIANA

Louisiana
Superdome

Kennedy
Space Center

FLORIDA

Coral
Castle

0   100   200   300 miles

READER'S GUIDE

*"Architecture should, among other
things, fulfill man's belief in the
nobility of his existence," said Eero
Saarinen, the master architect who
designed the monumental Gateway
Arch. From the elegant spans of the
Brooklyn and Golden Gate bridges
to the towering concrete crest of the
Hoover Dam, tamer of the Colorado
River, this book showcases some of
our nation's most ambitious feats of
engineering. Part I presents 10 man-
made wonders (labeled in green on
this map). Part II, the Gazetteer,
provides photographs, maps, and
descriptions of 21 others (bulleted
in green). A legend explaining the
symbols for facilities and map features is on the front inside cover.*

Arecibo
Observatory

VIRGIN
ISLANDS

PUERTO
RICO

0   50   100 miles

BROOKLYN BRIDGE

*Perhaps the most beautiful bridge
ever built, this span is a triumph of
19th-century American know-how.*

High above the choppy East River, pedestrians tramp across a wooden boardwalk, set above four lanes of traffic that whiz along the two roadways suspended below. The sightseers gaze toward Manhattan, enthralled by the view of the city's skyline. Overhead soar giant granite towers with twin Gothic archways—ancient gateways to a modern metropolis. Waves of curving cables connect the towers to the Brooklyn and Manhattan banks of the river, and a spider's web of diagonal stays and vertical suspenders holds the roadway secure. Despite its filigree appearance, the bridge possesses a reassuring solidity and permanence. There are longer bridges, bridges that carry more traffic, and bridges that embody more technical innovations, but there is only one Brooklyn Bridge.

When the Brooklyn Bridge was opened to the public in May 1883, it was dubbed the People's Bridge and hailed as the Eighth Wonder of the World. Walt Whitman was inspired by its "grand obelisk-like towers." Another poet, Hart Crane, devoted the last 10 years of his life to composing an epic tribute to "the most beautiful Bridge of the world." Countless photographers and painters—among them Joseph Stella and Georgia O'Keeffe—have attempted to capture the spell of its graceful network of cables and towers. Scores of Hollywood films have lingered over the sight of the bridge with Manhattan's skyscrapers twinkling in the distance—a scene of quintessential urban romance.

AUDACIOUS PLAN

A bridge across the East River was first suggested in 1800, but the waterway, with its treacherous currents and movement of tides, presented a monumental engineering challenge. For decades a fleet of ferries had linked Manhattan with Brooklyn, but the crossing was rough and frequently interrupted by bad weather. In winter, ice sometimes jammed the rivers, preventing the armies of clerks, textile workers, and laborers who resided in Brooklyn from getting in to work. And the number of passengers continued to rise. By 1860 an estimated 32 million people were taking the ferries each year, and the need for a road link could no longer be ignored.

When it came to deciding who would build the bridge, Brooklyn's city fathers chose John Augustus Roebling, already widely regarded as one of the greatest engineers of his time. The audacity of his design and the scale of the project were staggering. The bridge would be twice as long as any other suspension bridge then in existence and would span a distance of 5,862 feet, from Chatham Street in Brooklyn to City Hall Park in Manhattan. Because of the East River's importance to shipping, the bridge would cross the river in a single span and its deck would be high enough to clear the masts of the sleek clippers that thronged the port.

Roebling wanted to build his bridge out of steel, a material that was then largely untried in construction. He calculated the bridge would need to be 85 feet wide, with two lanes for cable cars and four lanes for horse-drawn vehicles and foot traffic. The key to the construction process was the sinking of giant caissons, which would serve both as digging chambers and forms for the bridge's foundations, into the bottom of the East River. (Caissons are wooden boxes that are pumped full of compressed air to keep out water and seeping muck.) After the workers reached bedrock, the caissons would be filled with concrete and become the foundations for the towers. The bridge's river span—the portion between the towers—would be 1,600 feet long, the longest span the world had yet seen. Roebling planned to drape main cables over the top of the towers and secure the cables to massive onshore anchorages located at each end of the bridge. Steel ropes, called suspenders, would hang

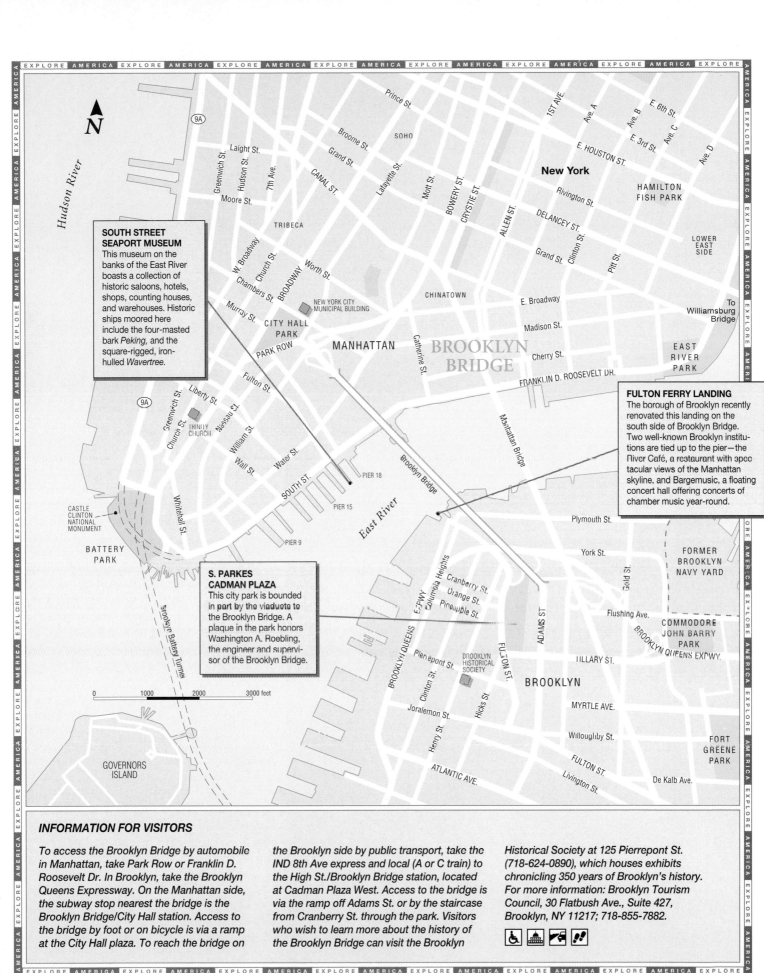

SOUTH STREET SEAPORT MUSEUM
This museum on the banks of the East River boasts a collection of historic saloons, hotels, shops, counting houses, and warehouses. Historic ships moored here include the four-masted bark *Peking,* and the square-rigged, iron-hulled *Wavertree.*

FULTON FERRY LANDING
The borough of Brooklyn recently renovated this landing on the south side of Brooklyn Bridge. Two well-known Brooklyn institutions are tied up to the pier—the River Café, a restaurant with spectacular views of the Manhattan skyline, and Bargemusic, a floating concert hall offering concerts of chamber music year-round.

S. PARKES CADMAN PLAZA
This city park is bounded in part by the viaduct to the Brooklyn Bridge. A plaque in the park honors Washington A. Roebling, the engineer and supervisor of the Brooklyn Bridge.

INFORMATION FOR VISITORS

To access the Brooklyn Bridge by automobile in Manhattan, take Park Row or Franklin D. Roosevelt Dr. In Brooklyn, take the Brooklyn Queens Expressway. On the Manhattan side, the subway stop nearest the bridge is the Brooklyn Bridge/City Hall station. Access to the bridge by foot or on bicycle is via a ramp at the City Hall plaza. To reach the bridge on the Brooklyn side by public transport, take the IND 8th Ave express and local (A or C train) to the High St./Brooklyn Bridge station, located at Cadman Plaza West. Access to the bridge is via the ramp off Adams St. or by the staircase from Cranberry St. through the park. Visitors who wish to learn more about the history of the Brooklyn Bridge can visit the Brooklyn Historical Society at 125 Pierrepont St. (718-624-0890), which houses exhibits chronicling 350 years of Brooklyn's history. For more information: Brooklyn Tourism Council, 30 Flatbush Ave., Suite 427, Brooklyn, NY 11217; 718-855-7882.

Sometime in 1882–83 a group of laborers on the Brooklyn Bridge, below, took time off from working on the new cable system to pose for the camera. Work on the bridge was carried on around the clock, primarily by immigrants. The building of the Brooklyn caisson required some 260 men each day. The work was so grueling that every week about one third of them quit and had to be replaced.

vertically from the main cables and support the roadway, or deck, of the bridge.

Roebling submitted his detailed plans for a bridge to Brooklyn's administrators in 1867. After the proposal was deemed feasible, a coalition of businessmen and legislators obtained the necessary state and federal authorization. The federal stamp of approval came after army engineers raised the headroom of the proposed bridge from 130 to 135 feet to ensure that the ships sailing through the harbor could pass under the bridge.

When the city fathers appointed John Roebling to be the bridge's designer, they could hardly have made a better choice. Born in Mühlhausen in what was then the German kingdom of Saxony, Roebling had trained both as an engineer and a bridge builder. He immigrated to the United States in

1831 at the age of 24, spent a short time farming in Pennsylvania, then took up work as a railroad surveyor and a designer of locks and canals. In 1841 Roebling developed a wire rope, or suspension cable, for use in the new Portage Railroad, a system built to haul canal boats over the Allegheny Mountains. He built a factory in Trenton, New Jersey, and began producing suspension cable on a large scale. Roebling's cable system proved successful on several suspension bridges he built in the 1850's and 1860's. One of these traversed the Niagara Gorge at Niagara Falls, a second one spanned the Allegheny River at Pittsburgh, and a third bridge crossed the Ohio River between Cincinnati, Ohio, and Covington, Kentucky.

TRAGIC
DEATH

It was the success of these difficult projects that convinced Brooklyn's civic boosters to entrust the bridge across the East River to Roebling. But tragedy came to pass: before work had even begun, the engineer was dead. While surveying the site for the tower on the Brooklyn end, Roebling was struck by a ferry as it was tying up at a dock; his right foot was crushed, requiring that the toes be amputated. After developing tetanus and suffering three weeks of agonizing convulsions, he died on July 22, 1869.

The project trustees' choice of a suitable successor fell naturally to the only man who could complete the project: Roebling's son, Col. Washington A. Roebling, an engineer and bridge builder who had served in the Union Army during the Civil War. Washington had assisted his father on the

CABLE, SUSPENDERS, AND STAYS
One of the Brooklyn Bridge's elegant arches, left, is barely visible through the tangle of cables, suspenders, and stays. The bridge's four cables are made of bundles of galvanized, high-strength steel wire that bear the weight of the steel suspenders supporting the deck. Each cable can take the weight of 380 suspenders. Diagonal stays radiate from the towers. All the stays were replaced beginning in 1987 after one of them became corroded and broke, killing a pedestrian.

Pittsburgh and Cincinnati bridges and had surveyed the site for the Brooklyn Bridge. He was familiar with the caissons that were essential to Roebling's design, having been sent to Europe by his father two years before to study them.

Today the bridge's caissons are buried beneath thousands of tons of stone. But the heroic efforts that put them there—a mix of technological triumph and human tragedy—live on as legends. The first caisson was launched off Brooklyn on March 19, 1870. The 3,000-ton chamber, measuring 168 feet long and 102 feet wide, was made of wood and iron, with a heavy roof, strong sides tapering to a narrow cutting edge, and no bottom. Once it sank to the floor of the river, the interior was filled with compressed air that forced the water out of the caisson. Soon teams of workers started to excavate, first digging with shovels and, when that proved too slow, blasting away with a newly invented smokeless powder explosive. The workforce, primarily Irish, German, and Italian laborers recruited from poor neighborhoods around the Brooklyn Navy Yard, entered and exited the chamber by means of airlocks. Separate spoil and water chutes were used to remove excavated material and maintain the required pressure.

A STROLL ACROSS THE BRIDGE
The elevated promenade, left, was designed by John Roebling to "allow people of leisure, and old and young invalids . . . to enjoy the beautiful views and the pure air . . . ." Reconstructed in 1982 for the bridge's centennial celebration, the promenade lies 18 feet above the center of the roadway. It is lined with benches and has a lane for cyclists, as well as a pedestrian walkway.

The towers of the Brooklyn Bridge, right, took five years to construct: the Brooklyn tower was completed in June 1875, and the New York tower was finished in July 1876. Both towers stand 276 feet 6 inches above the water line. Their arched openings are 33 feet 9 inches wide and 119 feet high.

The four main cables, below, that loop over the towers are 15.75 inches in diameter. Each cable is made from a bundle of 19 strands of steel wire, each strand comprising 278 steel wires.

Work inside the caissons was hot, arduous, and often dangerous. One of the most terrifying moments came on December 1, 1870, when workers discovered a fire in the massive timber roof caused by the blasting. Fed by compressed air, the fire had smoldered unnoticed through the Florida and Georgia pine. To extinguish the fire, Roebling flooded the caisson. The affected areas eventually had to be cut out and replaced by a team of 18 carpenters, who worked day and night for three months to repair the damage.

As the excavation continued, rocks were piled on the roof of the caisson, adding weight to help drive it deeper into the riverbed. On the Brooklyn side, progress was slow: huge boulders had to be broken up and removed so that the caisson could descend at an even level. On the Manhattan side, the site selected for the caisson turned out to be the location of one of the city's oldest garbage dumps, and workers had to remove tons of foul-smelling refuse before they reached the river bottom. The Brooklyn caisson hit bedrock at a depth of 44 feet 6 inches below high water, but the

Manhattan caisson, which measured 102 feet by 172 feet with a height of 21 feet 6 inches, descended to an astounding 78 feet 6 inches below high water before Roebling ordered the work to stop.

MYSTERIOUS MALADY

Roebling was forced to halt the sinking of the Manhattan caisson when an alarming illness began to plague his workforce. As they emerged from the airlocks after their shifts in the caisson, men began to complain of agonizing cramps, joint swellings, and partial paralysis. A crisis loomed when the first death was attributed to what was dubbed "caisson disease." Worse still, Washington Roebling himself fell victim to the mysterious malady.

Roebling and his men were suffering from decompression sickness, or "the bends," caused by a too-rapid transition from the high-pressure atmosphere inside the caisson to normal atmospheric

to supervise the project from his bedroom in Brooklyn Heights, writing detailed instructions for his hand-picked crew of assistant engineers. When Roebling's handwriting deteriorated, he relied on his devoted wife, Emily, who took down his instructions and relayed them to his subordinates.

Once the caissons were in position and filled with concrete, the task of finishing the towers went ahead. The towers were built of limestone up to the waterline, and of Maine granite above it. The four steel suspension cables would sit on four iron saddle plates mounted inside the top of each tower. The Brooklyn tower was finished in June 1875, and the Manhattan tower just over a year later.

As the towers rose above the East River, work crews began to construct the anchorages—one on the Manhattan side of the bridge and the other on the Brooklyn side. The anchorages would secure the cables that hold up the roadway as it crosses the river. The onshore anchorages are masonry

A view of Brooklyn from the Manhattan side of the East River, below, shows the Brooklyn Bridge and the Manhattan Bridge on the far left. The Manhattan Bridge (1909) was the third bridge to be built across the East River; the Brooklyn Bridge was completed in 1883 and the Williamsburg Bridge was finished in 1903. The Manhattan Bridge is 1470 feet in length and has a classically designed ceremonial approach, arch, and colonnade. Paved roads carry vehicular traffic and a lower deck transports subway trains across the river.

pressure at the surface. When the workers emerged after spending long hours in the compressed air of the airlocks, nitrogen bubbles suddenly formed in their bloodstreams, blocking the flow of oxygen to the body's tissues. Caisson disease claimed the lives of three men. Not until many years later would physicians identify it as the bends, a diagnosis derived from the observation of deep-sea divers.

Ravaged by caisson disease and the effects of overwork, Washington Roebling was forced to give up active work on the bridge. Though he would never again set foot on his creation, he continued

structures weighing 60,000 tons each and measuring the size of a city block; they are built around a cast-iron anchor plate, one to anchor each cable. Heavy iron eyebars arranged in a carefully calculated manner arc up to the top of the structure to the point where the bridge cables are attached.

Former sailors, accustomed to high rigging, were hired to spin the four cables—two outer cables and two near the middle of the bridge floor. Each cable describes a shape that is known as a catenary curve—the perfect natural form assumed by any rope or cable when it is suspended from two points.

COLOSSAL JOB

It took the workers more than a year to complete the spinning of the wires. The pivotal tool was the heavy steel "working rope," or traveler, fitted with a wheel that would unspool the cable wire back and forth between the Brooklyn and Manhattan anchorages. Starting on the Brooklyn side, workers secured a spool of wire on a drum and attached one end of the wire to the wheel. As it was pulled along the working rope by a steam winch, the wire traveled to the opposite anchorage, where it was looped around a steel shoe. As each spool ran out, workers spliced on the next

that saw the bridge as the road to riches, whether through control of bridge company stock, speculation in land, or the supply of materials for the construction. The administration was largely cleaned up in 1874, when the bridge was declared a project in the public interest. In June 1878, however, it was discovered that the company supplying the cable wire had been furnishing wire previously rejected by inspectors. By that time, hundreds of miles of substandard wire had already been spun into cables and could not be unraveled. Nevertheless, Washington Roebling opted to continue with the work: his father had incorporated

A ROOM WITH A VIEW
An office on the 24th floor of the New York City Municipal Building in Manhattan offers a splendid prospect of the Brooklyn Bridge and the East River.

wire using a small galvanized-steel ferrule threaded at both ends. At intervals along the cables, workers clamped wires into a cylindrical form and applied bindings to hold the strands together until the clamping and wrapping could be done. Each strand was a single continuous wire 185 miles in length. When 19 strands were completed, they were clamped, wrapped, and painted, becoming a cable consisting of more than 3,515 miles of wire.

The cable spinning brought to light some shady business dealings. Indeed, scandal dogged the construction of the bridge right from the start: there were powerful interests on both sides of the river

such a high margin of safety into his original design that the rejected wire would not compromise the bridge's stability in the slightest.

Once the cables were completed, vertical suspenders were attached to them and work began on the deck, a deep, hollow truss that was put together by attaching each section to a suspender and bolting it to its neighboring section. To stiffen the deck, John Roebling's design used stays that slanted down from the towers, angled across the suspenders, and were secured to the deck.

The Brooklyn Bridge took 14 years to build and cost the then staggering sum of $15 million. From

John A. Roebling wrote of the great towers, left, that he planned to build, saying that they "will serve as landmarks to adjoining cities, and they will be entitled to be ranked as national monuments."

Every day more than 121,000 trucks and automobiles travel across the Brooklyn Bridge, below, and some 1,000 people walk or bicycle along its wooden promenade. On its first full day of operation, 150,300 pedestrians crossed the bridge on foot, while 1,800 vehicles made the trip across the East River. Trains began running in September 1883, and, by 1888, they carried 30 million passengers annually.

the very beginning the idea of an enormous suspension bridge over the East River seized the public imagination. When the bridge was formally opened on May 24, 1883, New Yorkers enjoyed the biggest celebration they had seen since the Erie Canal was opened in 1825. The mayor of Brooklyn proclaimed an official holiday, and offices, shops, and factories on both sides of the East River emptied as workers flocked to witness the opening ceremony, presided over by Pres. Chester Arthur, who called the structure "a durable monument of democracy." At dusk the bridge's electric lights were switched on—the first use of electric light over a river—and a gigantic fireworks display lit up the night sky.

Although New Yorkers had much to celebrate, the building of the bridge had taken its toll. It is estimated that 27 people died during its construction. Seven days after the bridge opened, 12 people were trampled to death when throngs of sightseers were caught in a crush on the Manhattan side's promenade stairway and panicked.

In spite of this disaster, the bridge promenade became one of the most popular—and cheapest—spots from which to see the city. But many people feared that the bridge itself was unstable. To allay their fears, showman P. T. Barnum used his circus

know-how to prove that the bridge was safe: he led 21 elephants across it in May, 1884. Barnum needn't have bothered; more than a century after its construction, the Brooklyn Bridge has proved to require only minor maintenance, resurfacing of the roadway, and a periodic new coat of paint.

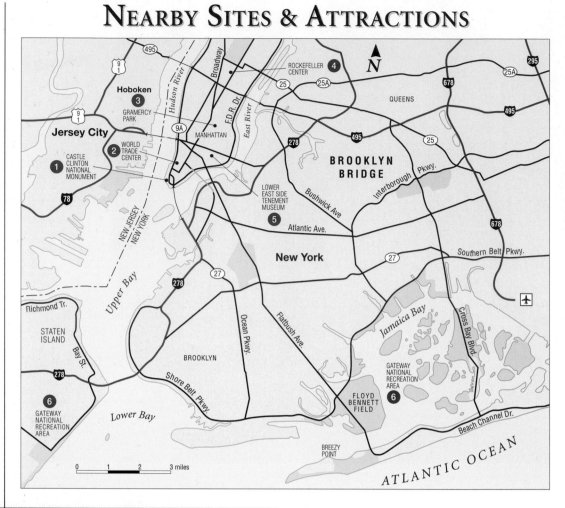

The eight-foot-thick walls of the Castle Clinton National Monument, below, are pierced with 28 gunports, which once held 32-pounder cannons. Originally called the Southwest Battery, the fort was renamed after the War of 1812 to honor former New York City mayor and state governor De Witt Clinton.

1 CASTLE CLINTON NATIONAL MONUMENT

This fort was built between 1807 and 1811 to protect New York Harbor in the event of war with Great Britain and France. When the sandstone structure was completed, it stood 200 yards offshore on an artificial island surrounded by 35-foot-deep water and was accessible from Manhattan by a causeway and drawbridge. The guns were never fired in war, however, and in 1824 the fort was rechristened Castle Garden and used as a recreation spot until 1855. By then landfill had replaced the causeway and, for the following 36 years, the island served as an immigration center—the first point of entry for millions of emigrants. In 1896 the fort was converted into the popular New York Aquarium, which was moved to Coney Island in 1941. The fort was on the verge of being demolished when World War II broke out; in 1950 it was declared a National Monument. A museum details the history of the fort. Located at 26 Wall St. in New York.

2 WORLD TRADE CENTER

Everything about this 10-acre complex is enormous: It is home to more than 500 businesses and trade organizations; more than 50,000 people work in the center; and an estimated 70,000 people visit it daily. The center's occupants generate 50 tons of garbage and consume 2.25 million gallons of water daily. Its imposing Twin Towers are the fourth-tallest buildings in the world. Each tower rises 1,350 feet (110 stories), has 21,800 windows, and houses 104 elevators. Every floor encompasses an entire acre of rental space. Numerous shops, restaurants, and banks are found within the center, as well as a police station and a medical center. An enclosed observation deck on the 107th floor and a rooftop observation platform

12 feet above the 110th floor provide magnificent views of New York Harbor, New Jersey, and Long Island. Located on Church St. in Manhattan.

3 GRAMERCY PARK

When New York lawyer Samuel Bulkley Ruggles sold his 66 building lots bordering on this park in 1831, he stipulated that only the lot owners could have access to the green space. Almost two centuries later, Gramercy Park retains its residents-only exclusivity. The public, however, can admire the Greek Revival homes surrounding the park, some of which display their original mid-19th-century cast-iron verandas. The Players, a Gothic Revival edifice nearby, was bought by actor Edwin Booth and turned into an actors' club after it was remodeled by New York architect Stanford White in 1888. A statue of Booth playing Hamlet stands in the center of the park and is visible through a gate at the south end of the square. Located on East 20th St. between 3rd Ave. and Park Ave. South in Manhattan.

4 ROCKEFELLER CENTER

The world's largest privately owned business center occupies 19 buildings spread over 22 acres of land. More than 240,000 workers and visitors pass through the complex daily. The center, which encompasses theaters, shops, and a skating rink, includes the Art Deco–style Radio City Music Hall, the country's largest theater, with 6,882 seats; the 850-foot-high RCA Building; and NBC Studios, which offers tours of its sets and control rooms. The statues, murals, furniture, and lighting fixtures within the complex

evoke the 1930's. Many buildings were built during the Depression and were financed by John D. Rockefeller Jr. Located on 5th and 6th avenues between 48th and 51st streets in Manhattan.

5 LOWER EAST SIDE TENEMENT MUSEUM

In 1929 a city bylaw stipulating that toilets be installed in every New York City apartment was instituted. Rather than comply, the owner of 97 Orchard St., an 1863 tenement building, kept the ground-floor and sub-basement storefront open, evicted the residential tenants, and had their 20 flats sealed up. For more than 50 years the units remained closed until they were reopened as a museum in 1988. Today visitors are guided through a cramped, narrow passageway and up a steep wooden staircase leading to several of the tiny, furnished 325-square-foot flats. One flat has uneven wooden floors, peeling wallpaper, and windows cut into interior walls. It is estimated that 10,000 people from more than 20 countries lived in these apartments between 1863 and 1935. The museum also offers walking tours of the working-class neighborhood and shows videos and slide shows on the American immigrant experience. Located at 90 Orchard St. in Manhattan.

6 GATEWAY NATIONAL RECREATION AREA

Established in 1972 and made up of three units, this park preserves the nation's oldest military site and lighthouse, as well as historic forts and airfields. It also features extensive beach areas and a wildlife refuge. Visitors can wander the landscape of the Staten Island unit—a mix of dunes, upland grasses, and freshwater wetlands—and tour Fort Wadsworth, which dates back to the Revolutionary War. The Breezy Point/Jamaica Bay unit encompasses the Jamaica Bay Wildlife Refuge, where more than 300 species of birds rest during spring and fall migrations. The refuge has an extensive hiking-trail system; the site's headquarters are located at Floyd Bennett Field in the Jamaica Bay unit.

A resident enjoys the privacy of Gramercy Park, above. Only the owners of the homes surrounding the park are allowed access.

The 1933 RCA Building, left, had the largest floor area of any commercial building in the world when it was erected. Specially designed setbacks make the 850-foot-high structure, which is the centerpiece of the Rockefeller Center, appear even taller.

New York's Skyscrapers

The Woolworth, Chrysler, and Empire State buildings are three stars of New York's skyline.

The date was April 24, 1913; the time precisely 7:29 p.m. Eight hundred guests sat silently in a banquet hall on the 27th floor of Manhattan's new 60-story Woolworth Building, waiting as a Western Union telegrapher tapped out a message to the White House that all was ready. In the nation's capital, Pres. Woodrow Wilson, like a magician waving a magic wand, pressed a button that closed a circuit and lit 80,000 electric lights that flashed for the very first time from every floor of the world's tallest skyscraper. The Woolworth Building was officially opened.

"A skyscraper," said Cass Gilbert, the architect of the Woolworth Building, "is a machine that makes the land pay." And on an island the size of Manhattan—12.5 miles long by 2.5 miles wide—the best way to make use of the land was to build upward. By the end of the 1920's, the island-city boasted 188 skyscrapers of more than 20 stories. Then, and now, visitors to the city streets found themselves strolling through a veritable outdoor architectural museum.

WORLD-FAMOUS SKYLINE

Overleaf: As night envelops the city, some of its landmark buildings light up the sky. The crown of the Chrysler Building still holds court amid its rivals, while the illuminated Empire State Building can be seen from miles away.

SELF-MADE MAN

The Woolworth Building was conceived by Frank Winfield Woolworth, who, at the age of 61, was in a financial position to plunk down $13.5 million (in 1913 dollars) in cold cash to build what was dubbed his Cathedral of Commerce. Just like his great edifice, which rose floor by floor to its final height of 792 feet, Frank's life had unfolded in stages. At 21 he had fulfilled his ambition of leaving the farm. He found work as a shop clerk in Watertown, New York, and began to formulate his own ideas about the retail trade. With money borrowed from his landlady, he opened his first store in Utica, New York, in 1879 and revolutionized merchandising by displaying his stock—everything from miniature tea sets, Christmas tree decorations, and goldfish to prayer books, wallets, and ice skates—where people could examine it. Each item was plainly marked with either a 5- or 10-cent label. Undaunted by the failure of his first store (blamed on its poor location), Frank opened another one in Lancaster, Pennsylvania, in 1879, and never looked back.

In 1912 Woolworth, nicknamed the Chief, consolidated his empire with that of five other retailers—four of them also sons of New York farmers, the fifth the son of a Connecticut blacksmith. Woolworth was now in a position to fulfill another great dream—to erect a magnificent building at Broadway and Park Place that would serve as the headquarters for the F. W. Woolworth Company.

Influenced by the architecture he had seen on his European travels, the Chief visualized the struc-

INFORMATION FOR VISITORS

The Woolworth Building is located at 233 Broadway Ave. between Barclay St. and Park Place. To reach it by subway, take the N and R line to the City Hall stop, the 2 and 3 lines to Park Place, or the 4, 5, and 6 lines to the Brooklyn Bridge station. The M1 and M6 buses on Broadway also provide access to the Woolworth Building. The lobby is open to the public during working hours. The Chrysler Building is located at 405 Lexington Ave. between 42nd and 43rd streets. By subway, take the 4, 5, 6, and 7 lines or the 42nd St. Shuttle to Grand Central Station. The M42 and M104 buses along 42nd St. also provide access to the Chrysler Building. The lobby is open to the public during working hours. The Empire State Building is located between 33rd and 34th streets at 350 Fifth Ave. If traveling by subway, take the 1, 2, 3, and 9 trains, the A, C, and E trains, or the B, D, F, N, R, and Q trains to the 34th St. station. By bus, take the M2, 3, 4, or 5 along Broadway or 5th Ave., or the 16 and 34 on 34th St. The observatories, located on the 86th and 102nd floors, are open daily from 9:30 a.m. to 11:30 p.m.
For more information: Woolworth Building, 233 Broadway Ave., New York, NY 10279-0003; 212-553-2000. Chrysler Building, 405 Lexington Ave., 31st Floor, New York, NY 10174; 212-682-3070. Empire State Building, 350 Fifth Ave., New York, NY 10118; 212-736-3100.

CONSTRUCTION SITE

An aerial view, right, shows the early stages in the construction of the Woolworth Building. The massive circular forms, surrounded by steel and wooden supports, are the caissons that, once completed, served as the pillars upon which the entire weight of the building relied.

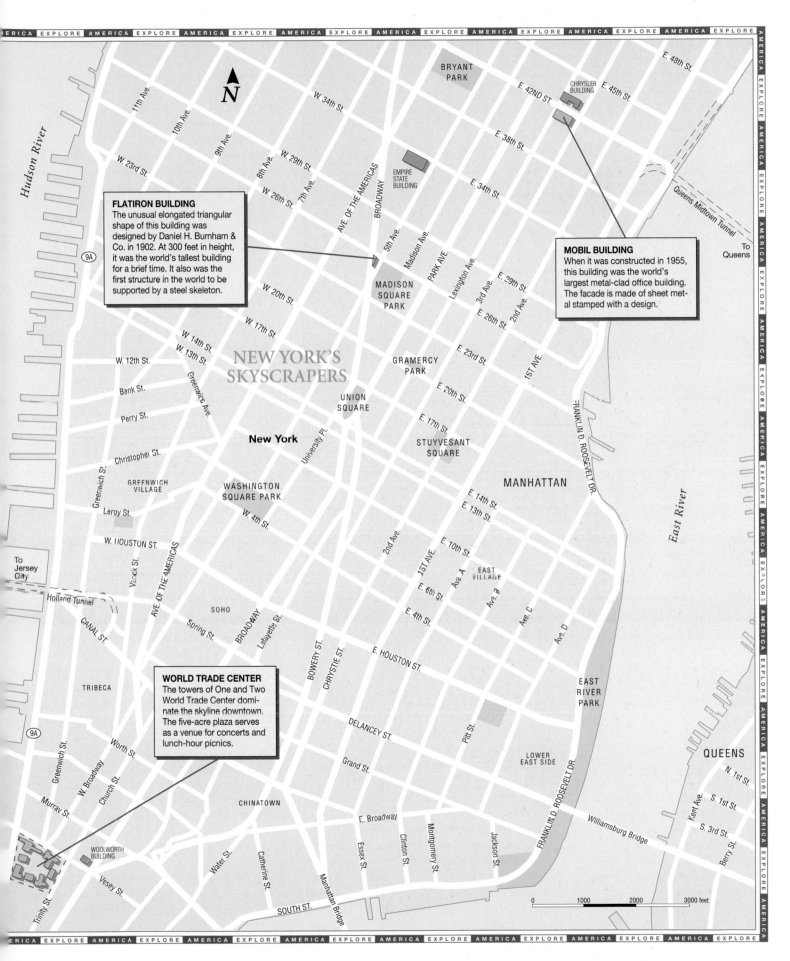

N

Hudson River

W. 34th St.

BRYANT PARK

CHRYSLER BUILDING

E. 42ND ST.

E. 48th St.

E. 45th St.

E. 38th St.

11th Ave.

10th Ave.

9th Ave.

W. 29th St.

W. 23rd St.

W. 26th St.

8th Ave.

7th Ave.

EMPIRE STATE BUILDING

E. 34th St.

Queens Midtown Tunnel

To Queens

9A

FLATIRON BUILDING
The unusual elongated triangular shape of this building was designed by Daniel H. Burnham & Co. in 1902. At 300 feet in height, it was the world's tallest building for a brief time. It also was the first structure in the world to be supported by a steel skeleton.

AVE. OF THE AMERICAS

BROADWAY

5th Ave.

Madison Ave.

PARK AVE.

Lexington Ave.

3rd Ave.

2nd Ave.

1ST AVE.

E. 29th St.

E. 26th St.

E. 23rd St.

MOBIL BUILDING
When it was constructed in 1955, this building was the world's largest metal-clad office building. The facade is made of sheet metal stamped with a design.

W. 20th St.

MADISON SQUARE PARK

GRAMERCY PARK

E. 20th St.

W. 17th St.

NEW YORK'S SKYSCRAPERS

W. 14th St.

W. 13th St.

W. 12th St.

Bank St.

Perry St.

Greenwich Ave.

UNION SQUARE

University Pl.

New York

E. 17th St.

STUYVESANT SQUARE

FRANKLIN D. ROOSEVELT DR.

East River

Christopher St.

Greenwich St.

GREENWICH VILLAGE

WASHINGTON SQUARE PARK

W. 4th St.

MANHATTAN

Leroy St.

W. HOUSTON ST.

Varick St.

AVE. OF THE AMERICAS

2nd Ave.

1ST AVE.

E. 14th St.

E. 13th St.

E. 10th St.

Ave. A

EAST VILLAGE

Ave. B

Ave. C

Ave. D

To Jersey City

Holland Tunnel

CANAL ST.

SOHO

Spring St.

BROADWAY

Lafayette St.

E. 6th St.

E. 4th St.

E. HOUSTON ST.

EAST RIVER PARK

9A

WORLD TRADE CENTER
The towers of One and Two World Trade Center dominate the skyline downtown. The five-acre plaza serves as a venue for concerts and lunch-hour picnics.

TRIBECA

BOWERY ST.

CHRYSTIE ST.

DELANCEY ST.

Pitt St.

LOWER EAST SIDE

Grand St.

Greenwich St.

W. Broadway

Worth St.

Church St.

WOOLWORTH BUILDING

CHINATOWN

Water St.

Spring St.

Essex St.

Clinton St.

Montgomery St.

Jackson St.

QUEENS

Murray St.

Vesey St.

Trinity St.

Catherine St.

E. Broadway

Manhattan Bridge

Williamsburg Bridge

Kent Ave.

N. 1st St.

S. 1st St.

S. 3rd St.

Berry St.

FRANKLIN D. ROOSEVELT DR.

SOUTH ST.

0   1000   2000   3000 feet

ERICA EXPLORE AMERICA EXPLORE AMERICA EXPLORE AMERICA EXPLORE AMERICA EXPLORE AMERICA EXPLORE AMERICA EXPLORE AMERICA EXPLORE

NEW YORK'S SKYSCRAPERS 23

ornate crown. Although it stands some 300 feet higher than the Pyramid of Cheops in Egypt, Gilbert's masterpiece appears lithe and graceful.

The interior is an outstanding work of art. The three-story lobby is a cross-shaped arcade with a vaulted domed ceiling that shimmers with glass tile mosaics similar in design to those found in the Byzantine churches in Ravenna, Italy. Light fixtures, cleverly concealed behind wrought-iron cornices faced in gold leaf, reflect soft light off the tiles, drawing visitors' eyes upward as if they were in a great cathedral. The walls of the lobby are lined with miles of golden-veined marble imported from the Isle of Skyros, off the coast of Greece. Bas-relief figures depict Woolworth counting his nickels, Gilbert peering through a pince-nez at a model of the building, and Gunwald Aus, the structural engineer, measuring a girder. Wide marble staircases wind gracefully to second-floor balconies adorned with frescoes and fittingly dedicated to "Labor" and "Commerce." Woolworth did not skimp: more marble was used for the wainscoting and terrazzo floors in the public corridors. And although the building could have had 79 stories if each floor had been 10 feet high, he insisted on more extravagant proportions: no ceiling was less than 12 feet high and some were more than 20 feet.

HEAVY-DUTY ENGINEERING The 15th-century appearance of the building's exterior belies its 20th-century structure. Sixty-nine caissons, each about 19 feet in diameter, were sunk 110 feet into the mud until they reached solid bedrock; the caissons were filled with concrete to form the piers that support the steel structure. Since the estimated weight of the building is 223,000 tons, each pier is carrying a load of about 24 tons per square foot.

The skeleton required some 24,000 tons of structural steel. Girders and columns were fastened together with rivets, which were inserted into predrilled holes. The rivets, resembling a bolt with a head at one end, were heated in a forge of fiery coke until red-hot. Workers used three-foot-long tongs to pluck them from the flames and toss them—sometimes as far as 50 to 70 feet—into a tin can. Other workers, also using tongs, picked up the rivets and fitted them into the aligned holes in the steel members. Yard-long riveting hammers drove the rivets home and flattened their heads. The rivets contracted as they cooled, tightening their grip on the steel. By November 15, 1911, the steel skeleton was visible above the pavement and by July 1, 1912, the Stars and Stripes flew atop the tower, anointing the steel frame.

When completed, the Woolworth Building enclosed 30 acres of floor space and had more than

QUEEN OF THE SKYLINE
An aerial view of the Woolworth Building, above, highlights its graceful tower, decorated with intricate terra-cotta detailing. In 1977 the Woolworth Corporation began a major restoration of the facade that took four years to complete. The exterior was carefully cleaned and its terra-cotta cladding was restored to its original splendor.

ture as a Gothic monument, similar in style to that of the Houses of Parliament in London, England. When Cass Gilbert asked how high the building should be, the Chief replied, "750 feet." "Am I limited to that?" asked Gilbert. "That's the minimum," his boss explained.

The Chief got his building, an elegant jewel of Gothic lines and delicate proportions trimmed in cream-colored terra-cotta tracery. Decorating it are whimsical gargoyles of bats, frogs, and owls, and statues representing Europe, Africa, Asia, and America. The building's base is a 29-story, U-shaped structure topped by a soaring 400-foot tower that begins at the 29th floor and culminates in an

If the Woolworth Building is a Gothic masterpiece, then the Chrysler Building sets the standard for Art Deco, the style that grew out of the 1925 *Exposition Internationale des Arts Decoratifs et Industriels Modernes,* held in Paris, France. The mastermind behind this building was another self-made man. Walter Percy Chrysler (1875–1940) began his career at the age of 17 working in a machine shop for five cents an hour. When he retired in 1935, he was the president of an automobile empire, the Chrysler Corporation. In 1928 Chrysler wrote that he "came to the conclusion that what my boys [sons] ought to have was something to be responsible for . . . . so the idea of putting up a building was born. Something that I had seen in Paris recurred to me. I said to the architects, 'Make this building higher than the Eiffel Tower.'"

Chrysler hired architect William Van Alen to design a building that would eclipse the Woolworth Building in height and at the same time be a

WOOLWORTH'S TREASURE
The intricate terra-cotta filigree shown at left surrounds the main entrance to the Woolworth Building. Cass Gilbert, the architect of this classic Gothic monument, including its elegant crown, below, said that the Gothic style allowed him the "possibility of expressing the greatest degree of aspiration . . . the ultimate note of the mass gradually gaining in spirituality the higher it mounts."

3,000 windows. The building required some 17 million bricks; 7,500 tons of terra-cotta; 28,000 tons of tiles; 53,000 pounds of bronze and iron hardware; 87 miles of electric wiring; and enough steel to build a 10-mile-long elevated railway. It was the first structure in New York to have its own power plant—four huge engines capable of generating enough power to run a city of 50,000 people. As a precaution against fire, a sub-basement pump that was able to deliver 500 gallons of water a minute to the top floors was installed.

F. W. Woolworth Co. moved into its new headquarters in late April 1913, and the Chief installed himself in a suite on the 24th floor. His office's cream walls were offset by gold and bronze furnishings. Floor and wainscoting in marble added another touch of luxury. His Empire-style desk was made of solid mahogany with bronze fittings, and a gold clock on the fireplace mantel, purchased by Woolworth in Paris, is believed to have been given to Napoleon Bonaparte by the czar of Russia.

The Woolworth Building reigned as the tallest building on earth until 1930 and showed the world just how far a small-town boy with a big dream could rise. On Woolworth's death in 1919, the *New York Sun* obituary penned a perfect portrait of the man: "He won a fortune, not in showing how little could be sold for much, but how much could be sold for little."

Each of the 30 doors of the passenger elevators in the Chrysler Building, right, bears a distinct Art Deco design made in metal and inlaid wood veneers. The fan-shaped lotus pattern is a hallmark of the Art Deco style. The interiors of the elevators are also lavishly decorated with exotic woods, embellished with metal detailing. Even the ceiling fans were fashioned in abstract designs.

AUTOMOBILE MOTIF

A winged hubcap, located on the 31st floor of the Chrysler Building, below, is one of the symbols of the automobile on the exterior of the Chrysler Building. A fanciful frieze of white and gray bricks resembling a row of automobiles is visible to the left of the winged gargoyle.

symbol of the Chrysler Corporation. The ground-breaking ceremony took place on October 15, 1927, and the building was officially opened to tenants on April 1, 1930. Throughout the construction phase, Chrysler faced serious competition from two other buildings that threatened to steal the coveted height record: the Bank of Manhattan Building at 40 Wall Street was going to be 925 feet, or 71 stories, and the Empire State Building was said to be reaching for 1,000 feet. To beat out his competitors, Van Alen added a feature that raised the building's height from 808 feet, or 68 stories, to 1,046 feet, making it the tallest structure in the world—higher than the 984-foot Eiffel Tower. Van Alen's ace in the hole was a 185-foot, 37-ton, stainless steel latticework spire that was raised into position after the edifice was completed.

Erecting the Chrysler Building had been a mammoth task: it required 21,000 tons of structural steel, 2,400 laborers, 391,381 rivets, 3,826,000 bricks, 3,850 plate glass windows, 35 miles of plumbing piping, and 750 miles of electrical wire. The structure consists of five buildings placed atop one another, wedding-cake style: the first building block rises to the 12th floor, the second from the 13th to the 25th, the third from the 26th to the 44th, the fourth from the 45th to the 57th, and the final section from the 58th to the 77th.

Chrysler's products provided the building with a leitmotiv. A frieze of automobiles, made of white and gray bricks, wraps around the facade just above the 26th floor. Eagle wings—the distinctive 1929 Chrysler automobile hood ornament—decorate the 31st story, and colossal gargoyles in the shape of eagle heads project from the corners of the 61st floor. Radiator caps, each 12 feet in diameter and shaped in the likeness of the Greek god Mercury, are found on some of the exterior cornices.

But it is above the 61st floor that the personality of the Chrysler Building is truly revealed. Its crowning glory is a 14-story terraced dome that tapers as it reaches for the clouds. Designed as an interlocking series of crescent-shaped arches pierced with triangular windows, the dome's sunburst motif still glitters and shines and is just as discernible amid the jumble of skyscrapers today as it was in 1930. The dome is sheathed in almost 48 tons of silvery, nontarnishing Nirosta, a blend of iron, chromium, and nickel. Chrysler chose the alloy because it retains its luster and color without extensive maintenance. The nails, bolts, nuts, and rivets in the dome are made of the same metal.

AUTOMOBILE SHOWROOM

Although the observation deck is no longer open to the public, visitors can examine the lobby, used as a showroom for Chrysler cars in the 1930's. The marble walls range in color from deep red to beige and are offset by a floor of Sienna travertine, assembled in patterns that point the way to the elevators and to the exits. Graceful, fanlike designs adorn the 30 elevator doors, each one constructed in wood veneers — bird's-eye maple, ebony, Japanese hardwood, plum pudding, myrtle burl, Oriental walnut—inlaid with steel. The ceiling of the lobby is covered by a large mural by Edward Trumbull (1884–1968), one of the most renowned American muralists of the time. Painted on canvas and cemented to the ceiling, the work measures about 110 feet by 97 feet and is believed to be the largest mural in the world. It depicts gigantic figures of muscular workers, the natural forces of fire, water, and lightning, ocean liners, railroad trains, airplanes, and dirigibles, including the *Graf Zeppelin*.

The Cloud Club, which once occupied the 66th to 68th floors of the building, is now just a fleeting memory of a bygone era. Opened during Prohibition, the club was the exclusive domain of executives in the automobile, aviation, steel, and oil industries. The main dining room was located within the first of the domes that form the building's crown. Its vaulted ceiling is decorated with a mural of clouds floating above the city's skyline. Fittingly, Chrysler's private dining room is ringed with a carved glass frieze of automobile workers. Although he had reached the pinnacle of his profession, Chrysler remembered his beginnings. When he began work as a machinist, he could not even afford to purchase his own tools, so he made them. What is more, he was both humble and proud enough to display those tools in a glass case for the public who visited the observation deck.

As the Chrysler Building neared completion, the "chase up into the sky," as the *New York Times* referred to it, was far from over. In August of 1929,

two months before the crash of the stock market, plans were already afoot to erect a building that would surpass the Chrysler Building in height. It would be called the Empire State Building, and one of the men behind its creation was John J. Raskob, the founder of General Motors and a rival of Chrysler's. Along with Raskob, the corporation included Coleman du Pont, Louis G. Kaufman, and Ellis P. Earle. Alfred E. Smith, the former governor of New York and a presidential candidate, was chosen to head up the corporation. The architectural firm of Shreve, Lamb & Harmon went to work on the concept and returned to the drawing board 16 times before "plan K" was accepted—the blueprint for the famous pencil-like edifice that now stands in the heart of Manhattan.

DEMANDING SCHEDULE Architect William F. Lamb later described the firm's mandate as being "short enough: a fixed budget, as many stories and as much office space as possible, an exterior of limestone, and completion by May 1931, which meant only a year and six months from the beginning of sketches." Excavation began on January

The radiant spire of the Chrysler Building, below, catches the last rays of a summer sunset. One of New York's most familiar landmarks, this Art Deco masterpiece still retains a mystique that draws New Yorkers and visitors alike.

22, 1930, and construction of the steel skeleton started on St. Patrick's Day, 1930. The framework rose at a rate of four and a half stories a week. To save time, temporary cafeterias were installed on five floors so that workers would not have to waste precious moments by leaving the work site for lunch. During the peak of construction there were about 3,000 men at work: 225 carpenters, 290 bricklayers, 107 derrick operators, 285 steelworkers, 249 elevator installers, 105 electricians, and 192 plumbers, as well as several other trade specialists. The building was completed on April 11, 1931—an astonishing five months ahead of schedule and $9 million under the projected budget of $50 million.

THE EMPIRE STRIKES BACK

The consortium had announced to the press that the building would be about 1,000 feet high, a deliberately vague figure that left room to build higher if competition appeared on the horizon. When the Chrysler Building's spire suddenly pierced the sky, Raskob and his colleagues realized that they needed to add to the height of the Empire State Building. Raskob's solution was to erect a dirigible mooring mast to the 86th floor, thus boosting the height by another 200 feet. At the time, transatlantic travel by such dirigibles as the famed *Graf Zeppelin* was popular. Raskob banked on the hope that airship travel would continue to grow and that his mooring mast would provide dirigibles with a sky-high dock. Once a dirigible had hooked onto the mast, an enclosed gangplank would be attached to the airship and passengers would alight and enter New York through the Empire State Building. This romantic notion, however, failed to take into account the wind factor at such a height. Only two

TAKING A BREAK
A couple of surefooted workers, above, take a break from polishing one of the gargoyles on the Chrysler Building. From this perspective, part of the Empire State Building is visible beneath the menacing face of the gargoyle.

MIRRORLIKE MASTERPIECE
The sunburst motif of the Chrysler Building's spire, right, shows off the intricate work that went into its construction. The six-level spire is made of an alloy called Nirosta, a rustproof, low-maintenance material that retains its luster.

private landings were attempted before the idea was abandoned for safety reasons. Still, the mast remains as the tip of the pencil-shaped tower.

The Empire State Building was officially christened on May 1, 1931, and not to be outdone by the opening ceremony of the Woolworth Building, Alfred E. Smith requested that Pres. Herbert Hoover flick a switch in Washington, D.C., that in turn powered all the lights in the building. The Empire State Building was now open for business, and throngs of sightseers flocked to see it, anxious to take a high-speed elevator ride up to the observation decks located on the 86th and 102nd floors.

On the first day the building was opened to the public, some 5,000 people toured it; within a month, that figure had reached 50,000. Bausch & Lomb telescopes, installed at each of the four corners of the observation deck on the 86th floor, provided a 180-degree view of the city and 80 miles beyond on a clear day. A photographer was on hand to take pictures of visitors with the city as the backdrop. Those who had paid an extra fee could board a small elevator to the second observation deck on the 102nd floor—the transformed airship

passenger lounge. The two observatories soon became one of New York City's greatest tourist draws, and by the end of the first year, a total of 775,000 people had made the pilgrimage, providing a gross income of $875,000 in ticket, souvenir, and refreshment sales.

CELEBRATION AND SORROW In 1932, a year after its opening, a searchlight beacon announcing that Franklin D. Roosevelt had been elected to the White House became the first light to shine atop the building, heralding the structure's love affair with the limelight. In 1956 revolving beacons, called the Freedom Lights—a symbolic gesture of welcome to new Americans—were installed. The top 30 floors were set aglow in April 1964 to kick off the New York World's Fair; in 1976 the mighty tower was bathed in red, white, and blue to celebrate the American Bicentennial; and on October 12, 1977, blue and white lights signaled that the Yankees had won the World Series. An automated color-changing apparatus was installed in 1984, allowing the colored lights on the moor-

The simplicity and straightforward design of the Empire State Building, right, symbolizes the city of New York, as well as the Empire State itself. Plans for the construction of the building were announced in August 1929, a mere two months before the stock market crash that unleashed the Great Depression of the 1930's. The old Waldorf-Astoria Hotel was demolished to make room for this building.

A steelworker standing on top of a beam during the construction of the Empire State Building, below, points his finger at the top of the spire of the Chrysler Building.

ing mast to be turned on and off at the flick of a switch.

Many famous people, from movie stars and sports figures to foreign dignitaries and literary luminaries, have come to pay homage to this great building. When Alice Liddell Hargreaves, the inspiration for Alice of Lewis Carroll's *Alice's Adventures in Wonderland,* visited New York in 1932, she proclaimed that her most impressive memory of the city was her trip to the Empire State Building. The actress Fay Wray, who was plucked from a mock-up tower by a great ape in the film classic *King Kong,* was surprisingly apprehensive when she took a trip to the top of the real building. Nevertheless, the movie was an instant success and projected the image of the Empire State Building onto the silver screen in theaters around the world. Fact, however, is stranger than fiction. On a foggy Saturday morning in July 1945, a B-25 bomber pilot with 34 bomb-

ing missions under his belt, was flying his 10-ton plane over Manhattan's airspace. The fog was so thick that the pilot couldn't see the top of the Empire State Building, which he used as his point of reference. Disoriented, he decreased his altitude from 2,000 feet to 1,000 feet in a vain attempt to duck below the fog and get his bearings. After frantically zigzagging to avoid other skyscrapers, the pilot found himself on a direct collision course with the towering building. The plane crashed into the north wall between the 78th and 79th floors, ripping open a gigantic hole 18 feet wide and 20 feet high. The impact tore off the plane's wings, which landed on the roof of the fifth floor; a propeller lodged in the limestone facade. The fuel tanks ignited, sending searing flames as high as the 86th floor. The pilot, a fellow crew member, and a passenger died in the crash, as did 11 office workers on the 79th floor. Although the building suffered no structural damage, it took a year and approximately $1 million to repair the fire damage.

SYMBOL OF A CITY

The Empire State is no longer the highest building in the world, but it still attracts thousands of visitors each year. Among them is the New York Road Runner's Club, which holds an annual race in its stairwells: runners scramble up the 1,575 steps from the lobby to the 86th floor. From dusk to dawn the building casts its golden halo around the top of the mast. During holidays and special events, the building projects the celebratory mood of the rest of the city. Cloaked in different colors to match the event—from the green lights of St. Patrick's Day, the red of St. Valentine's Day, the white and yellow lights of Easter Week, the red, black, and green lights of Martin Luther King Jr. Day, to the red and yellow that illuminate the sky from Halloween to Thanksgiving—the Empire State Building symbolizes the spirit of a great city.

LIGHTING UP THE NIGHT SKY
At night the top of the Empire State Building glows with a multitude of hues, left. The colors of the lights are changed to reflect various traditional holidays and celebrations. During spring and fall bird migrations, the lights are turned off on foggy nights so that any light shining through the fog will not confuse the migrators and cause them to fly into the building.

NEARBY SITES & ATTRACTIONS

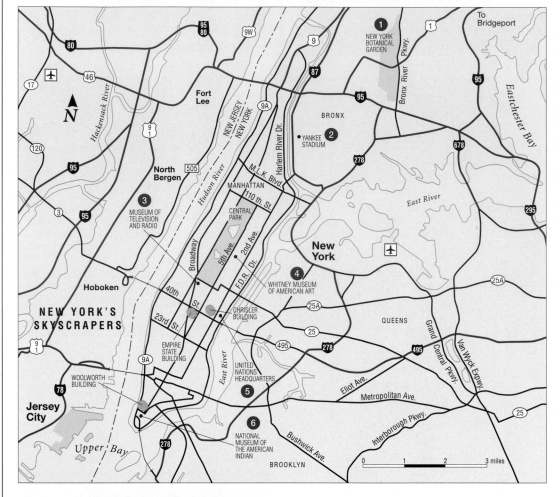

A New York landmark, the Enid A. Haupt Conservatory, below, reopened in May 1997 after four years of extensive restoration. The conservatory's permanent exhibit, titled "A World of Plants," features palms of the Americas, aquatic plants, tropical lowland and upland rain forest plants, and desert plants from Africa and the Americas. Special seasonal exhibitions are also held.

① NEW YORK BOTANICAL GARDEN

This 250-acre public garden was established by botanists Nathaniel and Elizabeth Britton in 1891. A portion of the 661-acre Lorillard family estate, bought by the city of New York in 1884, was set aside as a public garden by an act of the state legislature. Financial contributions were made by New York philanthropists, including Andrew Carnegie, J. Pierpont Morgan, and Cornelius Vanderbilt. The National Historic Landmark preserves wetlands, ponds, a waterfall, rock outcroppings, a 40-acre tract of virgin forest, 26 specialty gardens, and several historic structures. The Peggy Rockefeller Rose Garden displays more than 2,700 roses comprising over 230 species. The Native Plant Garden re-creates 11 different habitats found in the eastern United States. These include serpentine barrens, open meadows, coastal prairies, and woodlands. The 1902 Enid A. Haupt Conservatory—modeled after the conservatory at London's Royal Botanic Gardens and considered one of America's greatest examples of a Victorian glasshouse—holds 11 pavilions displaying tropical and desert plants. Visitors can see more than 11 varieties of palms and a 200-year-old skeleton of a saguaro cactus. In 1938 the city relocated several historic buildings to the grounds of the botanical garden. These include the 1840 Snuff Mill, which ground rose petal–flavored tobacco, a carriage house, and a stone cottage. The

34-room Lorillard family mansion is no longer on the premises: it burned to the ground in 1923. Located at 200th St. and Southern Blvd. in the Bronx.

2 YANKEE STADIUM

"The House That Ruth Built" is arguably the most famous site in the Bronx. The home of baseball's New York Yankees is a cropped field of rye and Kentucky bluegrass. The stadium seats more than 57,500 spectators and covers an area of almost 12 acres. When team owner Jacob Ruppert had it built in 1923, he specified that the distance between home plate and the right field fence be shortened to help Yankees' star Babe Ruth hit more home runs. On April 18, 1923, after playing at the Polo Grounds in Manhattan for nine years, the Yankees played their first game at Yankee Stadium, where Babe Ruth hit the stadium's first home run. The Yankees played their first night game here in 1946; today the playing field is lit by 800 multivapor and incandescent lamps. Statues and plaques highlighting the careers of famous Yankee players, managers, and owners are on display on the grounds. Located at 161st St. and River Ave. in the Bronx.

3 MUSEUM OF TELEVISION AND RADIO

This museum, founded by former CBS president William S. Paley, offers visitors the opportunity to enjoy more than 90,000 radio and television program tapes on 96 radio and television consoles. The museum also has a gallery that displays exhibits on the history of radio and television, and a theater that features a film on famous directors. Located at 25 West 52nd St. in Manhattan.

4 WHITNEY MUSEUM OF AMERICAN ART

The Whitney is the city's most important showcase for contemporary American art. Its founder, Gertrude Vanderbilt Whitney (granddaughter of railroad magnate Commodore Vanderbilt), was a patron of the arts and a sculptor in her own right. In 1907 she set up a studio in Greenwich Village where she exhibited the groundbreaking works of young American artists. In 1929 she offered to donate her large art collection to the Metropolitan Museum of Art. When the traditionalist museum spurned her offer, Mrs. Whitney set up her own museum. Originally on 8th Street, the Whitney Museum has been moved twice; it is now housed in a building designed in 1966 by Bauhaus architect Marcel Breuer. The controversial building on Madison Avenue is cantilevered out like the steps of an inverted pyramid. Its permanent collection includes abstract works by Stuart Davis and Georgia O'Keeffe, as well as works of realism by George Bellows and Edward Hopper. It also holds significant collections by Alexander Calder, Willem de Kooning, Roy Lichtenstein, Claes Oldenburg, Jackson Pollock, Frank Stella, and numerous other American artists. The museum has three branches located elsewhere in the city and one in Stamford, Connecticut. Located at 945 Madison Ave. in Manhattan.

5 UNITED NATIONS HEADQUARTERS

When the United Nations held its first general assembly in postwar London in 1946, its members agreed to establish a permanent site in New York City on an 18-acre stretch of land donated by John D. Rockefeller and valued at $8.5 million. The site is considered an international territory. The mandate of the United Nations is to decide on issues relating to peace, justice, and the economic and social well-being of the member nations. The UN Headquarters is composed of several buildings, including the 550-foot-high, 39-story Secretariat and the curved General Assembly buildings. Guided tours, which leave the public lobby of the General Assembly every 30 minutes, take visitors into the General Assembly Hall, where 2,103 seats accommodate UN delegates, the media, and the public. Works of art by international artists are displayed throughout the complex. A sculpture by Carl Frederick Reutersward of a knotted gun, titled *Non-Violence,* stands in the plaza of the General Assembly Building. Henry Moore's bronze statue, *Reclining Figure: Hand,* graces the north side of the Secretariat Building. Located on Fifth Ave. at 46th St. in Manhattan.

6 NATIONAL MUSEUM OF THE AMERICAN INDIAN

This museum, which is part of the Smithsonian Institution, has the largest collection of Native North American and South American artifacts in the nation. Some of the museum's early acquisitions were donated by George Gustav Heye, who in 1897 was working as a railroad engineer in Arizona when he purchased an Apache shirt—the first of many Native American artifacts he acquired. In 1989 Heye's collection was sold to the Smithsonian. Artifacts on display include Inuit stone carvings, wood carvings from the American Northwest, painted hides from the North American Plains Indians, Mayan jade, and Andean gold, as well as a collection of pottery and baskets from the American Southwest. Located at One Bowling Green in Manhattan.

The statue Let Us Beat Swords Into Ploughshares, *above, was given to the United Nations by the former Soviet Union in 1959. It is one of many pieces of art found on the grounds of the UN Headquarters.*

The Beaux Arts–style Alexander Hamilton U.S. Custom House, below, is the home of the National Museum of the American Indian. The 44 columns are decorated with the head of Mercury—Roman god of merchandise and merchants.

C&O CANAL

*Axes, plows, and shovels dug
this 19th-century canal out of the
rocky Potomac River valley.*

They wrested a great chunk of the North American continent away from the mighty British Empire, these brash upstart Americans, and with the 1803 Louisiana Purchase they acquired yet another vast swath of the continent from the French. They believed in their manifest destiny, in their land's boundless wealth, and in their own ability to tame the wilderness. "The Americans arrived but as yesterday on the territory which they inhabit," observed the French political scientist Alexis de Tocqueville after his 1831–32 journey through the young republic, "and they have already changed the whole order of nature for their own advantage. They have joined the Hudson to the Mississippi and made the Atlantic Ocean communicate with the Gulf of Mexico." And how had these enterprising Americans effected this change? With canals.

Even before America's Thirteen Colonies had broken their ties to England and coalesced into a country, frontiersmen had already breached the Appalachian spine that runs from Canada to Alabama. On the far side of the mountains, they found fertile soil for growing grains and other

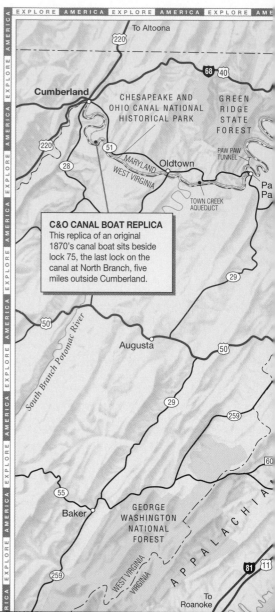

C&O CANAL BOAT REPLICA
This replica of an original 1870's canal boat sits beside lock 75, the last lock on the canal at North Branch, five miles outside Cumberland.

LONG-LOST LOCKS

Overleaf: The lock house at lock 6 in Montgomery County is one of several maintained by the National Park Service. Besides locking the canal boats through, a lockkeeper was responsible for maintaining and repairing the portion of canal under his care. Although most were in charge of only one lock, the lock-keeper living in lock house 49 was in charge of four locks—47 through 50—all within a quarter-mile distance and visible from his front door. Two of the locks at Four Locks, locks 47 and 48, above, are now a sea of grass.

produce and a wealth of timber and furs—but no convenient way to move these coveted riches to the enticing eastern markets, which were spread out along the Atlantic Coast.

To the visionaries of the new republic, canals were the answer to commerce and to nationhood. They would bind the country together and allow for the unlimited flow of valuable resources. "Rivers are ungovernable things . . . ," declared Benjamin Franklin. "Canals are quiet and very manageable." In Virginia, George Washington pressed hard for

SUNDAY RIDE

Riders cycle along the towpath, right, toward one of several bridges crossing the C&O Canal in Georgetown. This section of the canal was so busy in the 1870's that boats were lined up for about five miles, often having to wait four or five days to be cleared through.

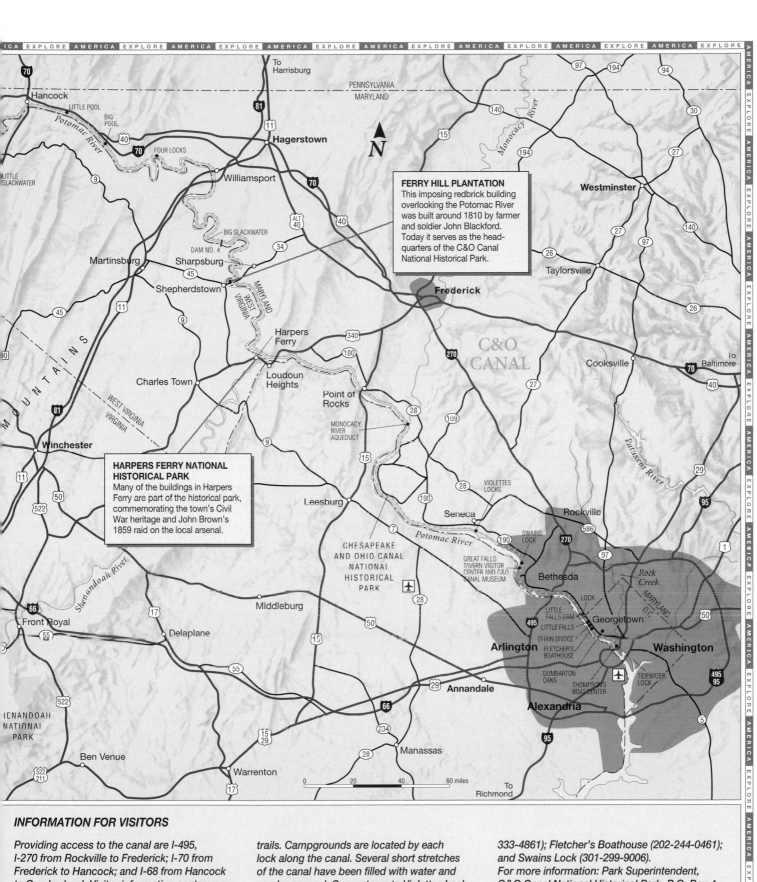

To Harrisburg

PENNSYLVANIA
MARYLAND

Hancock

LITTLE POOL

BIG POOL

Potomac River

FOUR LOCKS

Hagerstown

Williamsport

LITTLE SLACKWATER

ALT 40

BIG SLACKWATER

DAM NO. 4

Martinsburg

Sharpsburg

Shepherdstown

MARYLAND
WEST VIRGINIA

Harpers Ferry

Charles Town

Loudoun Heights

WEST VIRGINIA
VIRGINIA

Winchester

FERRY HILL PLANTATION
This imposing redbrick building overlooking the Potomac River was built around 1810 by farmer and soldier John Blackford. Today it serves as the head-quarters of the C&O Canal National Historical Park.

Westminster

Taylorsville

Frederick

C&O CANAL

Cooksville

To Baltimore

Point of Rocks

MONOCACY RIVER AQUEDUCT

HARPERS FERRY NATIONAL HISTORICAL PARK
Many of the buildings in Harpers Ferry are part of the historical park, commemorating the town's Civil War heritage and John Brown's 1859 raid on the local arsenal.

Leesburg

VIOLETTES LOCKS

Seneca

Rockville

SWAINS LOCK

Bethesda

Shenandoah River

CHESAPEAKE AND OHIO CANAL NATIONAL HISTORICAL PARK

GREAT FALLS TAVERN VISITOR CENTER AND C&O CANAL MUSEUM

Potomac River

LITTLE FALLS DAM
LITTLE FALLS
LOCK
Georgetown

Rock Creek

MARYLAND
D.C.

Patuxent River

Middleburg

Front Royal

Delaplane

Arlington

CHAIN BRIDGE
FLETCHER'S BOATHOUSE

DUMBARTON OAKS

Washington

THOMPSON'S BOAT CENTER

TIDEWATER LOCK

Annandale

ICNANDOAH NATIONAL PARK

Ben Venue

Warrenton

Manassas

Alexandria

0   20   40   60 miles

To Richmond

INFORMATION FOR VISITORS

Providing access to the canal are I-495, I-270 from Rockville to Frederick; I-70 from Frederick to Hancock; and I-68 from Hancock to Cumberland. Visitor information centers are located in Georgetown, Great Falls, Williamsport, Hancock, and Cumberland. Bicycles are permitted along most of the towpath, but not on park trails. Campgrounds are located by each lock along the canal. Several short stretches of the canal have been filled with water and can be canoed: Georgetown to Violettes Lock, Williamsport, Big Pool, Little Pool, Town Creek, and Old Town. Canoes, boats, and bicycles can be rented at Thompson's Boat Center (202-333-4861); Fletcher's Boathouse (202-244-0461); and Swains Lock (301-299-9006). For more information: Park Superintendent, C&O Canal National Historical Park, P.O. Box 4, Sharpsburg, MD 21782; 301-739-4200.

ICA EXPLORE AMERICA EXPLORE AMERICA EXPLORE AMERICA EXPLORE AMERICA EXPLORE AMERICA EXPLORE AMERICA EXPLORE AMERICA EXPLORE

C&O CANAL   37

the building of canals, becoming head of the newly formed Patowmack Company in 1785. Although the company's hopes for constructing canals that would skirt around falls, rapids, and other impediments to navigation never came to much, the promise of the Potomac River was far from dead. There was little question in most people's minds that the river was an obvious conduit for a grand canal from Chesapeake Bay to the Ohio River. Nor was there much doubt that the nation's new capital, now rising on the river's banks, was the logical terminus for such a canal.

Dreams for a canal along the Potomac were fueled by the success of New York State's 363-mile-long Erie Canal. Within a few years of that canal's opening in 1825, the states of Maryland, Virginia, and Pennsylvania chartered the Chesapeake and Ohio Canal Company to build a canal that would extend from Georgetown to the Ohio River, near Pittsburgh. Besides private investors, primary funding for the canal came from Washington, D.C., Georgetown, Alexandria, and Shepherdstown, Virginia (now West Virginia), as well as the state of Maryland. (Virginia later withdrew its funding once it discovered that the canal would be routed only along the Maryland side of the river.)

PORTENTOUS EVENT

On Independence Day, 1828, Pres. John Quincy Adams and an eager entourage boarded boats for a five-mile journey upriver to Little Falls. They disembarked, and the president proceeded with the speechifying and groundbreaking ceremonies that heralded the beginning of the Chesapeake and Ohio Canal (C&O). But the ground wouldn't break. Twice Adams tried to thrust his shovel into the dirt but hit a root. Finally, to the delight of the crowd, he took off his coat and jabbed again, this time with success. In hindsight, some observers would come to view the groundbreaking debacle as a portent of the problems that later plagued the star-crossed canal.

Already a problem was festering in Baltimore. Even as Adams struggled to pitch out the first spadeful of earth from the canal, proponents of the new "iron horse" broke ground for the Baltimore and Ohio Railroad (B&O). The two entities—canal versus railroad—would become fierce competitors, wrangling for decades over who would win land rights, business, and the people's trust.

In 1828, however, hopes were still running high among officials of the canal company. Though the C&O would cover only 184.5 miles to Cumberland,

TRADITION OF HOSPITALITY

The Great Falls Tavern, below, which serves as the main visitor center for the C&O Canal National Historical Park, first opened for business as a tavern in 1831. Originally named the Crommelin House in honor of a Dutch family that invested in the canal, the tavern offered hearty meals to travelers even after the canal closed in 1924. Today the museum displays historic photographs and artifacts and presents two short videos on the canal.

38

Maryland, making it the sixth-longest canal in the nation, its width of 60 feet at the water's surface and its depth of 6 feet made it the country's widest and deepest canal. Its enterprising creators, as author Walter S. Sanderlin wrote in his 1946 history of the canal, *The Great National Project,* were building for a prosperous future in which the canal would have "steamboats on its waters, industries along its banks and centuries of service ahead."

DAUNTING ELEVATION

After studying survey reports, the company directors decided that the canal would hug the northern bank of the Potomac River from the confluence of Rock Creek at the border of Georgetown and Washington, D.C., to the Maryland mountain town of Cumberland—a climb of 605 feet. Seventy-four stone-lined locks, averaging an increase of elevation of about eight feet each, would allow the canal to follow this route. But it would take a phenomenal amount of manpower wielding picks, plows, shovels, and stump pullers to bring this engineering vision to fruition.

The company awarded contracts to private builders who had bid on one-half-mile sections of the canal immediately above Little Falls. Most of the contractors had experience building canals in New York and Pennsylvania; like the C&O, they thought they knew their business. Yet difficulties immediately beset them. The soil above Georgetown proved as unyielding to their workers' spades as it had been to President Adams'.

Mostly gravel and hardpan, it had to be blasted out, and the blasting damaged local buildings.

Furthermore, labor problems plagued the canal builders incessantly. Because the Potomac Valley was largely rural, it had almost no labor pool from which to draw. And much of the itinerant manpower on the mid-Atlantic Coast was already tied up with harvesting or public works projects. Besides, it was not just pick-and-shovel men who were needed, but also experienced stonemasons and bricklayers—and they were in even shorter supply. Exacerbating the labor problem was the Potomac Valley's dread "sickly" season in summer, when the incidence of cholera increased as water levels fell to miasmic lows.

The obvious answer to this workforce dilemma was simply to recruit Europeans. Advertisements were placed in English and Irish newspapers. Some offers even included transportation, housing, food, whiskey, and $10 a month in exchange for a three-month indenture. With Great Britain in economic turmoil and unemployment high, men were ready to try their luck in America, and in the fall of 1829 hundreds of them packed into ships and crossed the Atlantic. What they found on the far shore depended on their contractor: some were treated decently, others were fed and housed poorly. In either case, the change in climate took a severe toll on their health, and the sick and homeless soon filled the poorhouses of Georgetown.

The canal also suffered from the poor quality of available raw materials. Though lime kilns had

sprung up along the river to supply cement, their product was inferior, as was the quality of the local stone needed to build the culverts, aqueducts, dams, waste weirs, locks, and lock houses.

But the worst blow came from the railroad company. The builders of the B&O Railroad had chosen the same route through the western mountain passes that the canal builders had mapped out. At places the route of the railroad and the canal would be so close to the mountains that there would not be room for both. Knowing this, the railroad had aggressively purchased the land first, then sought and won a court injunction that prohibited canal construction above Point of Rocks, which lay only 48 miles upriver from Georgetown.

Despite these difficulties, the canal project inched along. By 1831 it was open to the town of Seneca, about 23 miles into its 184.5-mile journey. The next year an appeals court ruled against the railroad and in favor of the canal, so work could proceed above Point of Rocks. But a new disaster was about to strike: in September a cholera epidemic spread south from Canada into the Potomac Valley. The workers, housed together in cramped and verminous dormitories, fell easy victim to the contagious disease. "Imagine the panic," wrote Charles Fisk Mercer, president of the canal company, in a report to his board, "produced by a man turning black and dying in twenty-four hours in the very room where his comrades are to sleep or to dine." Work came to a halt, and by year's end less than a dozen new miles of canal had been completed.

The lack of progress by 1833 pushed the canal company to the brink of insolvency, and the dream of the C&O seemed destined to die. Unwilling to let this happen, Maryland devoted more funds the following year to keep the project afloat. By the close of 1834 the canal was open to Dam No. 4, about 25 miles north of Harpers Ferry, and had snaked its way a total of 86 miles along the river even as land prices soared and labor unrest grew.

Boredom among the workers during the idle winter months helped breed trouble. As the years passed, rival Irish factions began sparring with each other and other European groups until a constant state of surreptitious warfare existed in some camps. Trying to quell the conflict, company officials finally expelled suspected troublemakers. While that eased tensions for a time, riots and bouts of violence still broke out periodically as the project dragged on through the 1830's and 1840's.

Unrest was particularly bad among the men working on the Paw Paw Tunnel. An engineering marvel in and of itself, the 3,118-foot-long tunnel was designed to avoid the Potomac's looping Paw Paw Bends and to reduce the canal route to one mile rather than the six miles of the natural river course. But the 27-foot-wide, 24.5-foot-high tunnel had to be bored through solid rock, and progress was tediously slow. The tunnel, begun in 1836, was not completed until 1850. During those years, work on the upper sections of the canal was frequently stalled by lack of funds and spring floods. Furthermore, contrary to initial figures that estimated that a crew of 40 workers could make it through 7 to 8 feet of rock per day, workers were capable of excavating only 10 to 12 feet per week.

<table>
<tr><td>INAUGURAL
TRIP</td><td>In 1850, 22 years and 6 presidential terms after Adams had broken ground for the C&O, workers completed the Paw</td></tr>
</table>

Paw Tunnel. On the morning of October 10, past problems and future worries were forgotten as crowds gathered to watch and wave as five canal boats "locked through" the rough-and-tumble town of Cumberland. According to a contemporary report in the Cumberland *Civilian* the boats were "laden with the rich product of the mines of Allegany and destined for Eastern Markets." Although one of the boats stopped in Williamsport and two got stuck above Dam No. 6 because their loads were too heavy, the remaining two boats made it to Georgetown in good time: seven days.

To get to Georgetown, the boats traveled through a total of 74 locks. Each lock was 100 feet long, 15 feet wide, and 16 feet deep, and was designed to raise or lower a boat about 8 feet. First, the boats were moved carefully through an upper gate into the lock. Once the boat was inside, the gate closed

and paddles (small iron sluice gates located at the bottom of the large wooden lock gates) in the lower gate opened, releasing the water in the lock. When the level of water in the lock became equal with that in the lower level of the canal, the lower gate would be opened and the boat drawn out. For the return journey up the canal, the process was reversed. Although the record was set at between 3 and 4 minutes, the average time for a boat to be locked through was about 10 minutes.

As well as traversing the locks, canal boats traveling between Georgetown and Cumberland crossed 11 aqueducts built above large streams. At Big Slackwater and Little Slackwater, where the Potomac was gentle and slow-moving, the boats actually followed the natural river course for a few miles. Although the route ran mostly through

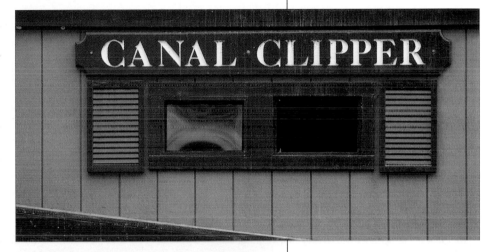

WINTER SCENE
Snow blankets a lock house along the C&O Canal, opposite page. During the winter a scow loaded with pig iron served as an ice-breaker on the canal until the ice became too thick. Some canallers lived on their tied-up boats through the cold months, but most moved into nearby towns, such as Williamsport and Shepherdstown.

RE-CREATING THE PAST
The C&O Canal National Historical Park operates two mule-drawn canal boats, the Canal Clipper, *above, out of Great Falls, and* The Georgetown, *out of Georgetown. Costumed interpreters, left, handle a boat and its mules and entertain passengers with canal lore. During the canal's heyday in the 1870's most captains owned two teams of two or three mules, which they worked in six-hour "tricks." While one team worked, the other rested in the boat's stable.*

A young musician relaxes by the towpath of the gently flowing C&O Canal, right. Despite recurring floods and underfunding, the towpath still follows the canal for all but 4 of its original 184.5 miles; the 4-mile gap has a well-marked detour around it.

One of the C&O Canal's original 74 locks is preserved in picturesque Georgetown, below. The neighborhood is the home of Georgetown University and the site of several excellent small museums, including Dumbarton Oaks, known for its lovely formal garden, created by landscape designer Beatrix Farrand, and its collections of Byzantine and pre-Columbian art.

countryside, the boats would occasionally pass towns that had sprung up in the frenzy of canal building and its promise of future commerce.

LIFE ALONG THE CANAL

The canal spawned several generations of canallers and lockkeepers who lived and raised their families with the income they earned working for the C&O. These hardy souls and their families lived in cabins at the rear of their boats, and their mules were stabled in a cabin up front. The mules did the hard work, walking along the towpath pulling tons of cargo behind them. This was not as arduous a task as might be expected: the canal was so well designed that, in the words of one canaller, "a boy ten years old could start the boat by making a long steady pull."

Lockkeepers were provided with a house, at least one acre of land for gardening, and no more than $50 a year to operate the lock. In return they had to be ready, day or night, to respond to the boatman's horn or his cry of "Hey-y-y-y-y Lock!"

In the 1870's the C&O Canal experienced its glory days: in 1875 alone it moved close to a million tons of coal, cement, flour, grain, and other produce. But success was short-lived and business soon slacked off. The railroad was faster, and less vulnerable to dry spells, floods, and winter freezes. Then, in 1889, the same devastating spring rains that caused the cataclysmic Johnstown flood in Pennsylvania also brought destruction to the canal, as torrents of floodwater virtually washed it away.

Bankrupted, the C&O Company had no resources with which to rebuild. In an odd twist of fate, the canal went into a receivership to its traditional nemesis—the B&O Railroad.

For three more decades the canal struggled on. As business declined, the railroad holding company that now owned the canal took much of the charm out of canal life. Now mules and boats were company-owned, and instead of bearing distinctive names like *Uncle Sam, Scow,* and *Jenny Lind,* the boats were simply numbered.

Today visitors flock to the 20,000-acre park to follow the towpath on foot, by bike—even on horseback. From the Georgetown locks, 22 miles of watered canal stretch almost to Seneca. At Great Falls, the Potomac rushes in a tumult past water-smoothed boulders and high rock cliffs. In contrast, the canal nearby maintains a mirror-smooth serenity, reflecting the hikers on the towpath.

Here and there, visitors pass by relics from the canal's past. Stone lockkeepers' houses provide a counterpoint to the wooded banks, and lovely old

MARYLAND STANDOFF
Sometimes canal boats would come head to head in the one-way channel inside the Paw Paw Tunnel, below. In theory the boat heading downstream had the right of way and the boat heading upstream was expected to back out. But more than once there was a standoff, with stubborn captains on both boats refusing to budge an inch.

In 1924, after spring floods tore away the C&O's locks and banks, the canal passed to the federal government, where it lay more or less dormant for years. But in 1954 there was talk of building a parkway, or landscaped road, along the canal. Associate Supreme Court Justice William O. Douglas challenged the editors of two pro-development papers to walk the towpath with him and see for themselves what a parkway would destroy. The walk resulted in the preservation of the canal, and in 1971 the canal and towpath became the Chesapeake and Ohio Canal National Historical Park.

towns like Harpers Ferry entice visitors to stop a while. From the graceful elegance of the Monocacy River Aqueduct, built to carry the waterway over the Monocacy River, to the weeping stones and black silence of the Paw Paw Tunnel, the park is marked by haunting reminders of a vanished dream. Sometimes on summer afternoons, as one of the park's canal boats makes a short nostalgic pull upstream, the illusion of the past is so strong that more than one traveler has imagined the sound of a boat horn summoning the lockkeeper to his task of helping the boat through the lock.

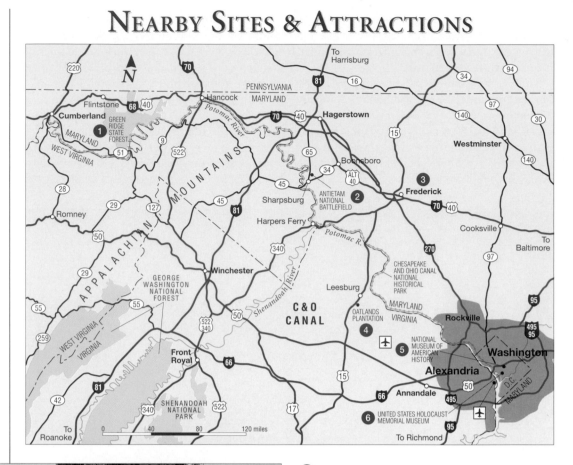

Parallel rows of weathered split-rail fences dissect Bloody Lane at Antietam National Battlefield, below. A monument to the 132nd Pennsylvania Regiment, which took part in this pivotal clash, overlooks the lane. An observation tower provides visitors with a view of the terrain where the battle was fought.

1 GREEN RIDGE STATE FOREST, MARYLAND

Encompassing approximately 40,000 acres of oak and hickory woodland, this is the second-largest state forest in Maryland. With the lowest annual precipitation in the state, the forest provides excellent opportunities for primitive and semiprimitive camping. Some campsites can be reached along the 24 miles of designated trails that zigzag along the ridges and stream valleys. Each spring many of the ponds and streams in the forest are stocked with trout. In the fall the forest is transformed into a brilliant palette of red, yellow, and orange foliage. Popular activities include hunting, horseback riding, mountain biking, bird watching, and off-road vehicle recreation. Located eight miles east of Flintstone off I-68.

2 ANTIETAM NATIONAL BATTLEFIELD, MARYLAND

On September 17, 1862, one of the bloodiest battles of the Civil War was waged at Antietam Creek, outside the town of Sharpsburg. In a bold attempt to push the war into the North, Gen. Robert E. Lee led 40,000 Confederate soldiers into Maryland. A copy of Lee's plans was intercepted and forwarded to Federal general George McClellan, who immediately mobilized 87,000 Union troops to thwart the invasion. The two armies clashed in three battles that fateful day. Lee's outnumbered forces were on the verge of being surrounded before reinforcements allowed them to retreat. In all, more than 22,000 men were killed or wounded during the bitter fighting, with 5,000 casualties sustained during the four-hour clash

WE ARE THE SHOES, WE ARE THE LAST WITNESSES.
WE ARE SHOES FROM GRANDCHILDREN AND GRANDFATHERS,
FROM PRAGUE, PARIS, AND AMSTERDAM,
AND BECAUSE WE ARE ONLY MADE OF FABRIC AND LEATHER
AND NOT OF BLOOD AND FLESH, EACH ONE OF US AVOIDED THE HELLFIRE.

MOSES SCHULSTEIN (1911—1981),
YIDDISH POET

on a sunken country road, later named Bloody Lane. Walking trails in the 12-mile battlefield are dotted with interpretive markers and maps that outline the pivotal events that took place here. From June through October costumed interpreters depict camp life for visitors. A visitor center is located north of Sharpsburg on Hwy. 65.

3 FREDERICK, MARYLAND

Frederick was established in 1745 by Daniel Dulany, a lawyer and land speculator from Annapolis. The scene of strife during the War for Independence and the Civil War, the town has preserved its historic buildings surprisingly well. The Courthouse Square has served as the town meeting place since the mid-1700's; it was here that inhabitants protested British taxation and where Abraham Lincoln spoke following the Battle of Antietam. The Barbara Fritchie House and Museum pays tribute to a staunch Federalist who, at the age of 95, defiantly waved the Stars and Stripes from her porch in the face of the advancing Confederate army. Located 50 miles west of Baltimore on Hwy. 70.

4 OATLANDS PLANTATION, VIRGINIA

This stately mansion was built in 1803 by George Carter, the great-grandson of Robert "King" Carter, a wealthy landowner. The public is welcome to tour many of the rooms in the Greek Revival home, which is decorated with turn-of-the-century furnishings. As the plantation flourished, more land was purchased

and, in its heyday, the 300-acre plantation also boasted a gristmill, brick factory, and a blacksmith's shop. Italian Renaissance gardens—a series of interconnecting terraces designed by George Carter—are open to the public. Located six miles south of Leesburg on Hwy. 15.

5 NATIONAL MUSEUM OF AMERICAN HISTORY, DISTRICT OF COLUMBIA

A storehouse of Americana, this branch of the Smithsonian Institution houses a diversity of historic items, from the tent that Gen. George Washington lived in during the Revolutionary War to car racing champion Don Garlits' sleek dragster *Swamp Rat XXX*. The vast collection includes displays as diverse as the flag that flew over Fort McHenry in 1814, a Foucault pendulum, a Model T Ford, and a lightbulb made by Thomas Edison himself. Other exhibits display farm machines, ship models, and antique automobiles and steam locomotives. Also here is an exhibit on the Crystal Palace, the site of the 1851 World's Fair in London, England. Located on Constitution Ave. between 12th and 14th streets.

6 UNITED STATES HOLOCAUST MEMORIAL MUSEUM, DISTRICT OF COLUMBIA

Dedicated to preserving the history of the Holocaust, this museum pays homage to the people who lost their lives during this horrific period. The victims included 6 million Jews, as well as millions of other people. When visitors enter the museum's permanent exhibit they are given an identity card containing the photograph and brief biography of one of the victims of the slaughter. Empty canisters of Zyklon-B, homemade toys, and a railcar used to transport Jews to the Nazi concentration camps are on display. Films, photographs, and eyewitness accounts recount many stories of heroism. Located at 100 Raoul Wallenberg Place in Washington, D.C.

A pile of shoes, left, belonging to people who died in Nazi concentration camps during World War II is on exhibit at the United States Holocaust Memorial Museum. The inscription reads, in part, ". . . because we are made of fabric and leather/And not of blood and flesh, each one of us avoided the hellfire."

The National Museum of American History, below, first opened its doors in 1964 as the Museum of Science and Technology. An authentic 1860's country store, relocated from Headsville, West Virginia, is on display in the main lobby.

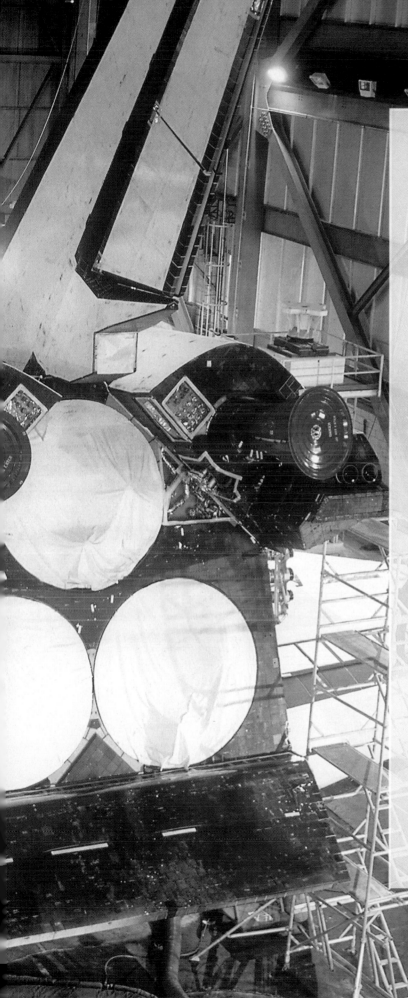

KENNEDY SPACE CENTER

Built to propel human beings into outer space, this complex is one of the most thrilling on earth.

t's early morning, and on an island punctuated by palmettos and scrub oaks, an alligator lazily soaks up the warmth of the Florida sun. Overhead a bald eagle floats majestically, the pinion feathers at the end of its wings straining to catch a thermal updraft as the bird arcs ever higher in the sky. Nearby, a round-tailed muskrat scurries by and darts into the salt grass.

Suddenly, a searing light blazes forth, illuminating the landscape like the rising of a second sun. A few seconds later comes the sound: a deep-throated, preternatural rumble that builds, second by second, into a deafening roar. Slowly at first and then picking up speed, the $2 billion space shuttle *Endeavour*—one of four such crafts in NASA's fleet—rises into the sky.

The shuttle soars higher and higher, flashing 600-foot-long flames from its two solid rocket motors. The land quivers from the buffeting blasts. Soon the lights from the boosters wink

With the Vehicle Assembly Building looming in the background, egrets and other birds, above, ply the wetlands that surround the Kennedy Space Center. The Merritt Island National Wildlife Refuge is home to many animals considered threatened or endangered, including the Eastern indigo snake and the West Indian manatee. A seven-mile drive in the refuge hooks up with a five-mile hiking trail, giving visitors a chance to view the refuge's flora and fauna.

Overleaf: The space shuttle Columbia *sits in the Vehicle Assembly Building prior to its launch on November 19, 1997— the 88th shuttle mission in NASA history.*

out as they finish expending their fuel. Six minutes later, traveling at nearly 10 times the speed of a bullet fired from a rifle, the shuttle orbiter arrives in earth's orbit powered by its three main engines. Only a wispy white trail of spent rocket fuel remains to mark its rumbling ascent into space. Meanwhile, the alligator opens its jaws, turns its head slightly, and lumbers away to search for its morning meal.

The incongruity seems almost surreal: the space shuttle, a supreme symbol of 20th-century technology, sharing the landscape with a reptilian creature whose prehistoric ancestors trod the earth's surface long before the dinosaur. But this is the Kennedy Space Center, a magical world of gargantuan buildings and equally impressive machines—and home to a wildlife refuge boasting the most diverse collection of threatened and endangered species in the continental United States.

For three decades the Kennedy Space Center has given birth to a host of triumphs and tragedies, legends and dreams. And still it endures, the most famous gateway to the one frontier that humans will never conquer or fully comprehend, no matter how long they survive on earth—space.

Though the names Kennedy Space Center and Cape Canaveral are often used interchangeably, the two are, in fact, different sites. Situated halfway along Florida's eastern shore, Cape Canaveral is an elbow-shaped strip of land that juts into the

Atlantic Ocean. The Kennedy Space Center is situated on Merritt Island, a mile or so to the northwest across the Banana River.

From 1961 to 1968, Cape Canaveral was the home base of all of America's manned spaceflights, beginning with the 15-minute suborbital hop of Alan Shepard, the first American in space, continuing through the one-man Mercury and two-man Gemini missions, and ending with the launch of *Apollo 7.* In the 1950's the cape had served as a missile test range for rockets with such names as *Bomarc, Snark,* and *Hound Dog*—relics that have since faded into both military and rocket history.

CHANGING VENUES

The growth of the Apollo project, which would eventually send man to the moon, and the development of the means to get there—the *Saturn V,* the largest rocket ever built—called for facilities that dwarfed anything Cape Canaveral had to offer. In 1962 work began on Merritt Island on Launch Complex 39—two launch pads and an array of support buildings. The Kennedy Space Center was born later that year and took over the role of America's launch pad for manned spaceflight in 1968 with the flight of *Apollo 8,* the first manned mission to leave earth's orbit and circle the moon. Meanwhile, adding to the public's confusion over the site's geography, the name Cape Canaveral had been changed to

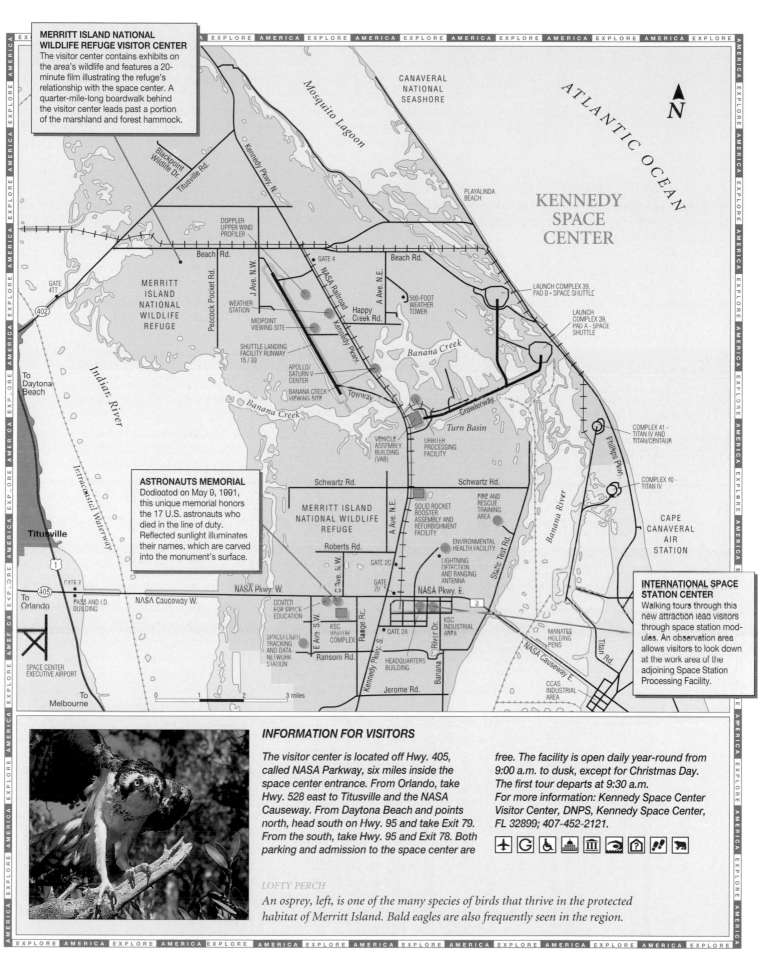

MERRITT ISLAND NATIONAL WILDLIFE REFUGE VISITOR CENTER
The visitor center contains exhibits on the area's wildlife and features a 20-minute film illustrating the refuge's relationship with the space center. A quarter-mile-long boardwalk behind the visitor center leads past a portion of the marshland and forest hammock.

ASTRONAUTS MEMORIAL
Dedicated on May 9, 1991, this unique memorial honors the 17 U.S. astronauts who died in the line of duty. Reflected sunlight illuminates their names, which are carved into the monument's surface.

INTERNATIONAL SPACE STATION CENTER
Walking tours through this new attraction lead visitors through space station modules. An observation area allows visitors to look down at the work area of the adjoining Space Station Processing Facility.

CANAVERAL NATIONAL SEASHORE

Mosquito Lagoon

ATLANTIC OCEAN

N

KENNEDY SPACE CENTER

PLAYALINDA BEACH

Blackpoint Wildlife Dr.

Kennedy Pkwy. N.

Titusville Rd.

DOPPLER UPPER WIND PROFILER

Beach Rd.

GATE 4

Beach Rd.

GATE 4TT

402

MERRITT ISLAND NATIONAL WILDLIFE REFUGE

Peacock Pocket Rd.

J Ave. N.W.

WEATHER STATION

MIDPOINT VIEWING SITE

NASA Railroad

Happy Creek Rd.

A Ave. N.E.

500-FOOT WEATHER TOWER

LAUNCH COMPLEX 39, PAD B - SPACE SHUTTLE

LAUNCH COMPLEX 39, PAD A - SPACE SHUTTLE

Indian River

SHUTTLE LANDING FACILITY RUNWAY 15 / 33

Kennedy Pkwy.

APOLLO/ SATURN V CENTER

Banana Creek

To Daytona Beach

BANANA CREEK VIEWING SITE

Towway

Banana Creek

Turn Basin

Crawlerway

COMPLEX 41 - TITAN IV AND TITAN/CENTAUR

Intracoastal Waterway

VEHICLE ASSEMBLY BUILDING (VAB)

ORBITER PROCESSING FACILITY

Phillips Pkwy.

COMPLEX 40 - TITAN IV

Titusville

Schwartz Rd.

Schwartz Rd.

MERRITT ISLAND NATIONAL WILDLIFE REFUGE

A Ave. N.E.

SOLID ROCKET BOOSTER ASSEMBLY AND REFURBISHMENT FACILITY

FIRE AND RESCUE TRAINING AREA

Banana River

CAPE CANAVERAL AIR STATION

Roberts Rd.

ENVIRONMENTAL HEALTH FACILITY

Static Test Rd.

1

GATE 2C

LIGHTNING DETECTION AND RANGING ANTENNA

405

GATE 3

C Ave. N.W.

GATE 2B

To Orlando

PASS AND I.D. BUILDING

NASA Pkwy. W.

NASA Pkwy. E.

NASA Causeway W.

Range Rd.

CENTER FOR SPACE EDUCATION

KSC VISITOR COMPLEX

KSC INDUSTRIAL AREA

3

NASA Causeway E

MANATEE HOLDING PENS

Titan Rd.

SPACE CENTER EXECUTIVE AIRPORT

SPACEFLIGHT TRACKING AND DATA NETWORK STATION

E Ave. S.W.

GATE 2A

River Dr.

HEADQUARTERS BUILDING

Banana

CCAS INDUSTRIAL AREA

To Melbourne

Ransom Rd.

Kennedy Pkwy. S.

Jerome Rd.

0   1   2   3 miles

INFORMATION FOR VISITORS

The visitor center is located off Hwy. 405, called NASA Parkway, six miles inside the space center entrance. From Orlando, take Hwy. 528 east to Titusville and the NASA Causeway. From Daytona Beach and points north, head south on Hwy. 95 and take Exit 79. From the south, take Hwy. 95 and Exit 78. Both parking and admission to the space center are free. The facility is open daily year-round from 9:00 a.m. to dusk, except for Christmas Day. The first tour departs at 9:30 a.m.
For more information: Kennedy Space Center Visitor Center, DNPS, Kennedy Space Center, FL 32899; 407-452-2121.

LOFTY PERCH

An osprey, left, is one of the many species of birds that thrive in the protected habitat of Merritt Island. Bald eagles are also frequently seen in the region.

Cape Kennedy in 1964 following President Kennedy's assassination the previous year. Three years later the name was switched back to Cape Canaveral after protests from local residents who fought to preserve its original name.

NASA officials quickly realized that the island that would become the astronauts' future launch site was already a treasure trove of plant and animal life. In 1963 they turned over the wildlife management to the Department of the Interior, which created the Merritt Island National Wildlife Refuge. Today all of the sanctuary's 140,000 acres lie within the Kennedy Space Center boundaries.

With a diverse habitat that ranges from sand dunes to pine forests, the refuge is a haven to 22 endangered and threatened species and more than 300 species of birds. Creatures ranging from ocean mice and indigo snakes to bobcats and

1,000-pound leatherback turtles all call the Kennedy Space Center home—not to mention 19 different species of mosquitoes. Curiously, the shuttle launches seem to have had no discernible effect on the region's wildlife, although refuge workers occasionally have to respond to NASA requests to shoo an alligator off the shuttle landing strip.

BUILT ON A LARGE SCALE

Some buildings at the site impress visitors because of their style or architectural elan. Others simply overwhelm by their sheer size and scope. The Kennedy Space Center's Vehicle Assembly Building, or VAB, fits squarely into the second category. From a

distance, the structure looks like a large box plunked down on the sun-baked savanna. But when visitors alight from a tour bus and stand next to it, the colossal size of the building invariably fills them with awe and wonder. Visible from the air from more than 50 miles away, the VAB towers 525 feet high and covers 8 acres.

The VAB's inner core, with its four high-bay assembly areas, is so cavernous that one myth contends it once boasted its own weather system. As the story goes, when the air-conditioned climate inside the building proved too much of a contrast with the hot, humid air outside, clouds formed inside the structure and caused rain to fall. NASA engineers say that while condensation on pipes and catwalks sometimes causes drops to fall, the reports of full-blown clouds indoors are entirely exaggerated.

The building's outer shell is covered by 1 million square feet of insulated aluminum panels. When a 200-foot-long American flag and bicentennial symbol were painted on the southern side of the building in 1976, NASA had to order 6,000 gallons of paint for the job. Attached to the eastern end of the building is the Launch Control Center and the Firing Rooms, where a raft of busy engineers and technicians monitors shuttle launches.

Built originally to assemble the various stages of the *Saturn V* into a ready-to-launch rocket, the VAB was modified in the late 1970's to perform a similar role for the space shuttle. That work starts when the four segments of each solid rocket motor are hoisted onto the Mobile Launch Platform, a two-story steel structure that serves as a portable launch pad. The segments are then joined to form two complete boosters. Next comes the bulbous external tank, which arrives by a barge in the nearby Turn Basin; the tank provides fuel for the shuttle's three main engines while also serving as the structural backbone of the assembly. Finally, the orbiter is towed into the VAB from the Orbiter Processing Facility a few hundred yards away, where it is serviced and checked out after each flight. Lifted into a vertical position by a 250-ton crane, the DC-9-sized craft is lowered onto the launch platform and bolted to the tank and boosters—an operation that demands tolerances of less than a hundredth of an inch.

Once the mating—or "buildup operation" in NASA parlance—is completed, the 456-foot-high doors on the east side of the building open up in preparation for the shuttle's short trip to the launch pad. But this is no ordinary delivery job: the huge vehicle that transports the shuttle is only slightly less awe-inspiring than the shuttle itself.

The giant crawler-transporter looks as though it has sprung straight from the Brobdingnagian world of *Gulliver's Travels*. The vehicle consists of four immense double-tractor treads with cleats that weigh 1 ton apiece. Driven by crews who work inside two cabs situated at diagonally opposite corners, the crawler begins its work by backing into the VAB under the shuttle. It then jacks up its 11-million-pound load six feet in the air before trundling off to the launch pad at a stately speed of one mile an hour. (The gas guzzler gets one mile per 150 gallons of diesel.) A laser docking system provides pinpoint accuracy for the final delivery.

ROAD TO
THE STARS

A vehicle such as the crawler demands a special type of road, and NASA engineers created a dandy: a highway as wide as an eight-lane freeway. But it is not so much what shows on the surface that is impressive—it is what lies underneath. To support the crushing

SOME ASSEMBLY REQUIRED
The Vehicle Assembly Building, above, is one of the world's largest buildings. Standing some 525 feet tall and stretching another 716 feet long, the VAB covers a ground area of almost 8 acres.

DIFFUSING THE DANGER
Situated below the Mobile Launch Platform, the flame trench, left, is an essential component of the launch pad. The curved shape of the deflector steers the huge flames generated at liftoff to the side of the trench. Without the trench the flames would shoot straight down, bounce back up and engulf the vehicle. Bus tours of the space center stop near both launch pads.

51

The space shuttle Discovery, *below, prepares for one of its launches. The orange external fuel tank carries the bulk of the fuel used to send the shuttle into orbit. It disengages about nine minutes into the flight.*

weight of the crawler, NASA engineers built the road a full seven feet thick. At the bottom lies three feet of limestone, covered by fill and asphalt and topped off with eight inches of river rock. As the crawler makes its snail's-pace trip to the launch pad, the weight of the vehicle reduces the top layer of the roadway to sand, creating better traction.

At the other end of the crawlerway lie two octagonal launch pads—39A and 39B—each a quarter square mile in size. Perched on top of a concrete mound that slopes up to level areas of 48 and 55 feet above sea level, respectively, each pad is an engineering marvel in its own right. Two trenches bisect the pads to deflect the flames and heat away from the shuttle's engines and boosters. Shaped like an inverted V, the trenches stretch almost 500 feet in length. Six 12-foot-high cylindrical spouts called rainbirds dot the pad. Working in conjunction with a series of pipes and nozzles, the rainbirds lay down a cushion of water for the few critical seconds before and after the launch; this protects the orbiter and its payload from the earth-shaking rattle produced by the solid rocket motors and the three main engines. The Sound Suppression System is fed by water from a nearby tank and can spew 300,000 gallons of water over the Mobile Launch Platform in about 20 seconds.

HISTORIC EVENTS Despite their utilitarian look of steel and cement, these launch pads have given birth to some of the most momentous events of the 20th century. None can top the morning of July 16, 1969, when *Apollo 11* lifted off into a hazy, humid Florida sky under the anxious watch of more than 1 million people stretched out on nearby beaches and highways—and more than 600 million around the world by television. Sucking 15 tons of fuel a second into its huge F1 engines, the craft rocketed toward a destination some quarter-million miles away. Four days later, the mission that began at Launch Pad 39A culminated when Neil Armstrong stepped onto the lunar surface in the Sea of Tranquillity—the first human being to walk on another world.

The last operational Saturn V rocket left the earth in 1973, sending America's first space station, *Skylab,* into orbit. Fortunately, visitors to the Kennedy Space Center can get a firsthand look at one of three surviving rockets at the Apollo/Saturn V building. Situated a half-mile from the VAB, the 100,000-square-foot complex is dominated by a Saturn V lying on its side, supported by huge steel trusses. The rocket stages are separated so that visitors can gaze at the inner workings of a rocket more than a football field in length, from the five first-stage engines to the tip of the escape tower that topped the three-man spacecraft. Nearby stands the lunar module, the ungainly looking craft that ferried the astronauts down to the lunar soil. The building also features a 1960's vintage Firing Room complete with actual launch consoles, status boards, and control blocks used during the heady days of the Apollo missions.

One of the space center's two crawlers, left, dwarfs a tour bus as the massive machine lumbers down a specially designed road carrying the 9-million-pound Mobile Launch Platform. The crawler is one of the largest tracked vehicles in the world—weighing in at about 6 million pounds. The surface of the crawler is equivalent in size to that of a baseball diamond.

DOING AWAY WITH SPLASH-DOWNS

In the Apollo days, astronauts returned to earth by splashing their module down in the Pacific Ocean to be picked up by an aircraft carrier. But when the Kennedy Space Center introduced the concept of round-trip shuttle flights to space travel, splashdowns became obsolete. Now, after firing the descent engines on the opposite side of the planet, the shuttle commander guides the orbiter back into the earth's atmosphere. Five minutes before touchdown, visitors standing on Merritt Island can watch as the craft, now flying at subsonic speed, swings over the Indian River and then arcs to the right or the left, depending on wind direction, for a landing on runway 15 or 33.

The landing looks like a regular aircraft's with one main exception: there is no noise. Lacking propulsion in the descent stage of the flight, the orbiter literally glides back to earth with no chance for the pilot to circle and try again.

Stretching for some 15,000 feet, the runway is about twice the length of the average commercial airport runway. It features a series of red lamps and white lights at different heights. When the pilot is on the correct final glide slope of 1.5 degrees, the runway's red and white lights appear superimposed over one another. A few seconds later, the tail wheels touch down at 226 miles an hour. Then the nose wheel eases onto the pavement and a parachute pops open to help bring the shuttle to a stop.

And so a journey that began 6 million miles ago comes to an end on Merritt Island, less than five miles away from where it all began at Launch Pad 39B. Even as the astronauts unbuckle their seat belts, work has already begun in the VAB, assembling another shuttle for the next mission—and the start of another adventure at the Kennedy Space Center, America's threshold to space.

Visitors explore the space center's Rocket Garden, below. Eight authentic rockets are displayed on the grounds, including a Mercury-Atlas rocket similar to the one that launched John Glenn into space.

Nearby Sites & Attractions

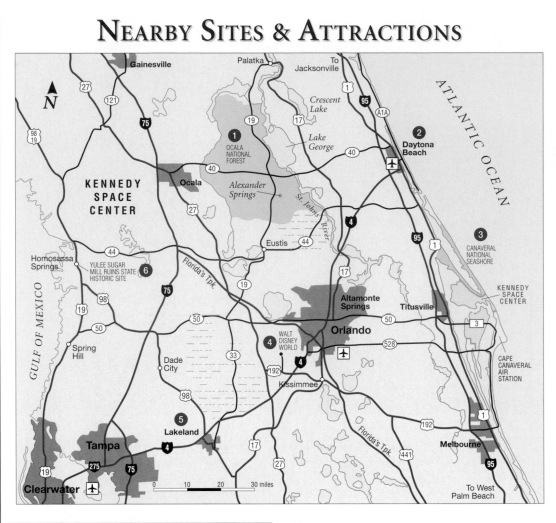

A pastel blue building, below, sits atop the pier at Daytona Beach. Not far from the pier—popular with local fishermen—a wide promenade leads visitors to an observation point called the Space Needle.

1 OCALA NATIONAL FOREST

This 383,000-acre subtropical forest encompasses a wildlife area, several natural springs, and the world's largest stand of sand pine. The forest's more than 200 ponds and lakes are popular with boaters, swimmers, and fishermen. Three natural springs feed 50 miles of pristine rivers; Alexander Springs alone produces 76 million gallons of water daily. Hiking trails, including a section of the Florida National Scenic Trail, weave through the thick forest of longleaf, cypress, and slash pines. The forest is home to one of the largest herds of deer in the state. Located on Hwys. 40 and 19.

2 DAYTONA BEACH

This city is renowned for the automobile races held here throughout the year. The hard-packed beach is 23 miles long and 500 feet wide at low tide, making it ideal for car racing, a tradition that began in the early 20th century. Thirteen automobile speed records were set at Daytona between 1902 and 1935, including Sir Malcolm Campbell's 1935 mark of 276 miles an hour. Cars are still permitted on the beach at low tide. The Daytona International Speedway hosts race car, stock car, and motorcycle races, including Speed Weeks throughout the year and the Daytona 500 in February; the Paul Revere 250 and the Pepsi 400 are in July. Elsewhere in the city, the Daytona Beach Arts and Sciences Museum displays American decorative arts, pre-Columbian artifacts, and Cuban and Floridian art. Located on Hwys. 1 and 95.

3 CANAVERAL NATIONAL SEASHORE

This 57,000-acre area of wilderness, beaches, lagoons, and dunes has been protected since 1975. The seashore preserves more than 1,000 species of plants, numerous endangered animal species, and 25 miles of waterfront—the longest stretch of unspoiled beach on Florida's east coast. More than 300 species of birds have been spotted here, including many that stop during their spring and fall migrations. Endangered or threatened birds such as Southern bald eagles, wood storks, scrub jays, and peregrine falcons frequent the area. Other endangered animals include West Indian manatees, Eastern indigo snakes, and loggerhead sea turtles, which bury their eggs on the beach in holes dug with their flippers. Two beaches, Playalinda and Apollo, are popular for swimming, boating, and fishing, and hiking trails wind through hummocks of old-growth oaks draped with Spanish moss. At the northern end of the seashore, Turtle Mound rises 50 feet. This hardened refuse heap of oyster shells, charcoal, and food scraps is said by archeologists to be evidence of a prehistoric civilization. The mound was used by Spanish explorers as a lookout and navigational beacon. Located two miles east of Titusville on Hwy. 3.

4 WALT DISNEY WORLD

This 28,000-acre complex includes a theme park, exhibition halls, golf courses, campgrounds, and resort hotels. The Magic Kingdom is based on Disney characters and stories, giving visitors young and old the chance to mingle with Caribbean pirates, visit a haunted house and Cinderella's castle, ride a mine tram, and tour the Hall of Presidents. They can even stop and hug Mickey Mouse and other Disney characters who roam through the theme park. Another favorite attraction, EPCOT Center, focuses on the achievements of science and industry. Located 20 miles southwest of Orlando off Hwy. 192.

5 LAKELAND

Surrounded by 13 lakes, this city in the heartland of the Florida citrus belt is a popular outdoor recreation area and cultural center. Lakeland hosts a number of annual events, including the Sun 'n Fun Fly, which exhibits old war aircraft and experimental airplanes. The Mayfaire-by-the-Lake Festival, held in May, attracts musicians, dancers, and artisans who perform in Elizabethan costume. The city is also home to Florida Southern College; its west campus buildings were designed by Frank Lloyd Wright. Lakeland is also the spring training headquarters of baseball's Detroit Tigers. On the cultural front, the Polk Museum of Art complements its permanent collection of European ceramics and pre-Columbian artworks with changing exhibits of contemporary and historic art. Located on Hwys. 4 and 98.

6 YULEE SUGAR MILL RUINS STATE HISTORIC SITE

This site preserves the ruins of a sugar mill that was operated by David Levy Yulee between 1851 and 1864. One of Florida's most compelling figures, Yulee was elected Florida's congressional territory delegate in 1841 and voted in as governor when Florida entered the Union in 1845. A member of the Congress of the Confederacy during the Civil War, Yulee used his mill to supply sugar for Southern troops. When Confederate leaders asked him to provide them with raw materials by destroying his Atlantic and Gulf Railway, Yulee refused. More than 100 slaves toiled on his 5,100-acre plantation until May 1864, when Union naval forces burned Yulee's house to the ground. A concrete path is dotted with numerous plaques that trace the history of the mill. Visitors can view the partially restored mill, with its native limestone chimney, and the remains of a grinding machine that was used to process raw sugar cane. Located in Homosassa Springs.

The turquoise water of Juniper Springs, above, is a tranquil oasis in Ocala National Forest. The water temperature of the spring hovers around the 72°F mark.

Thick forest surrounds the grinding machine, left, at the Yulee Sugar Mill Ruins State Historic Site. The mill's original owner, David Levy Yulee, was accused of aiding the escape of Jefferson Davis at the end of the Civil War. Although he was briefly imprisoned, Yulee was later set free under presidential orders by Ulysses S. Grant.

LOUISIANA SUPERDOME

*The dream of one man, this
massive stadium stands as a
monument to engineering genius.*

As airplanes fly over the Gulf of Mexico on their approach to New Orleans, they bank left on their course to the runway, offering passengers a clear view of a section of the city's downtown that lies strung along the silty and serpentine Mississippi River. In the streets below, visitors to the Big Easy are charmed by the myriad landmarks that make this most European of American cities unique: the ornate facades of the French Quarter, the mansions of the Garden District, Zydeco and Cajun music spilling from open windows, the streetcars and riverboats, and the dialects heard on the bustling wharves. But from above, the one feature visitors are likely to notice more than all the others is not a spired cathedral nor a gilded mansion—not even the Mississippi itself—but rather a massive copper bowl that appears to have been placed there by a giant hand: the Louisiana Superdome.

From the sky, the Superdome appears too large for its surroundings, much like the saucer-sized belt buckles that adorn the waists of the cowboys

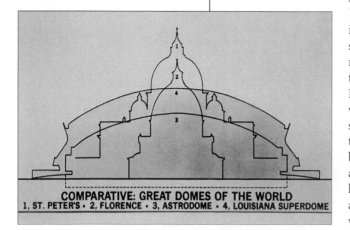

COMPARATIVE: GREAT DOMES OF THE WORLD
1, ST. PETER'S • 2, FLORENCE • 3, ASTRODOME • 4, LOUISIANA SUPERDOME

who once competed in rodeo championships inside. Immense and hulking, the Superdome seems to have devoured half the inner city.

In fact, this spectacular edifice has snuggled nicely into the body and soul of New Orleans and, far from intruding, has been credited with rescuing the city in the 1970's from a descent into decay. Originally conceived as little more than a modest 50,000-seat stadium to lure professional football to New Orleans, the Superdome grew into a bold shrine that would inspire more superlatives than Mardi Gras or the city's famed restaurants.

"The Louisiana Superdome will make all other stadiums in existence as obsolete as Rome's colosseum," gushed the *New York Times* as the largest indoor stadium ever built slowly began to rise out the reclaimed swamp. Halfway through its construction, Ben Levy, the man who would one day manage the stadium, looked up at the tapestry of steel girders being stitched into a 9.7-acre roof 27 floors above him, spread his arms wide, and cried out, "Go ahead, world, top this!"

When the stadium was completed in 1975, a *Chicago Tribune* reporter wrote: "Someday people everywhere will point with wonder and envy at what New Orleans has wrought." The *Houston Chronicle* declared that the complex "glistens like a giant dewdrop diamond on the throat of New Orleans," while the *Saturday Evening Post* trumpeted, "The State of Louisiana has come mighty close to erecting a Garden of Eden for the delectation of its citizens and visitors in New Orleans."

Starry-eyed hyperbole? Perhaps, but the Louisiana Superdome is so enormous—and its existence the result of such audacious engineering and political cunning (some called it chicanery)—that language is hard-pressed to contain it.

BIG DREAM, BIG STADIUM

Standing on the stadium floor, visitors feel minuscule, swallowed up by the dimensions of the structure. Beneath its roof, the arena is nearly half a mile in circumference and covers 125 million cubic feet of space—30 percent more than the Great Pyramid of Cheops in Egypt. Gazing up at the 273-foot dome, which is 680 feet in diameter, visitors have the illusion that they are looking at an overcast night sky.

In an age of computer-aided design, high-tech materials and equipment, and ambitious civic vision, it is not surprising that such a gigantic

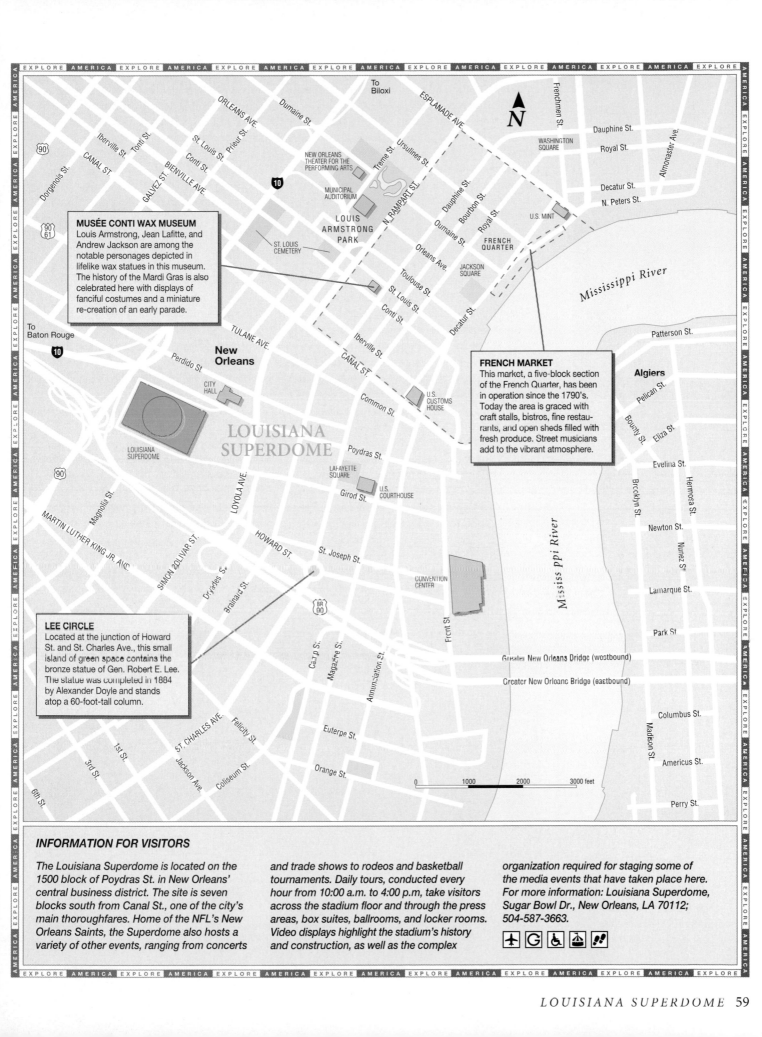

To Biloxi

N

MUSÉE CONTI WAX MUSEUM
Louis Armstrong, Jean Lafitte, and Andrew Jackson are among the notable personages depicted in lifelike wax statues in this museum. The history of the Mardi Gras is also celebrated here with displays of fanciful costumes and a miniature re-creation of an early parade.

New Orleans

FRENCH MARKET
This market, a five-block section of the French Quarter, has been in operation since the 1790's. Today the area is graced with craft stalls, bistros, fine restaurants, and open sheds filled with fresh produce. Street musicians add to the vibrant atmosphere.

Algiers

Mississippi River

To Baton Rouge

LOUISIANA SUPERDOME

Mississippi River

LEE CIRCLE
Located at the junction of Howard St. and St. Charles Ave., this small island of green space contains the bronze statue of Gen. Robert E. Lee. The statue was completed in 1884 by Alexander Doyle and stands atop a 60-foot-tall column.

0   1000   2000   3000 feet

INFORMATION FOR VISITORS

The Louisiana Superdome is located on the 1500 block of Poydras St. in New Orleans' central business district. The site is seven blocks south from Canal St., one of the city's main thoroughfares. Home of the NFL's New Orleans Saints, the Superdome also hosts a variety of other events, ranging from concerts and trade shows to rodeos and basketball tournaments. Daily tours, conducted every hour from 10:00 a.m. to 4:00 p.m., take visitors across the stadium floor and through the press areas, box suites, ballrooms, and locker rooms. Video displays highlight the stadium's history and construction, as well as the complex organization required for staging some of the media events that have taken place here. For more information: Louisiana Superdome, Sugar Bowl Dr., New Orleans, LA 70112; 504-587-3663.

The New England Patriots square off against the Green Bay Packers, above, in Super Bowl XXXI. The game, which was held in front of a Superdome crowd of more than 72,000 on January 26, 1997, marked the fifth time the National Football League title match was played in the Superdome. Green Bay prevailed 35–21 to win its third NFL championship.

IMPRESSIVE STATISTICS

The steel skeleton of the stadium rises from the ground during early construction, right. In all, more than 20,000 tons of structural steel were used to build the Superdome. Some of the stadium's other staggering statistics include more than 19,000 light fixtures, 550,000 feet of aluminum siding, and some 400 miles of interior electrical wiring.

building could be built. Still, everything topographic, climatic, and economic argued against its being built in New Orleans.

There were grave doubts whether the spongy, unstable soil of New Orleans, a city that sits four feet below sea level, could support a 300,000-ton structure. There were also concerns about a building of such monumental girth withstanding the hurricanes that prowled the coast, or whether the runoff from the roof would be too much for the machinery designed to pump rainwater from the low-lying city. But the main stumbling block was financial: the city lacked the money for a project whose price tag would soar to $163 million. Only with the state's help could the ambitious project be funded.

Given Louisiana's rancorous political history and the capabilities of the construction business in the 1960's, when the dome was still years away from the drafting table, a casual observer would have deemed the prospects murky at best. But such obstacles did not deter Dave Dixon, a businessman who had been dreaming of attracting a National Football League team to New Orleans since the late 1950's. Along with his interest in pro football, Dixon was tantalized by the architectural musings of Buckminster Fuller. Fuller believed that in the future cities would be covered by huge domes. Combining his two visions, Dixon began enlisting supporters for the realization of his dream.

In the late 1960's Dixon found an eager ally in Gov. John McKeithen, an impassioned football fan and, more important, a persuasive orator. At no small political risk, McKeithen persuaded the Louisiana legislature to let voters decide the stadium's fate by a referendum. Days before the vote, the National Football League promised a franchise to New Orleans, contingent upon a new stadium. The pro-Superdome side won by a landslide.

Despite the victory, opposition persisted and various lawsuits challenged the project. Critics and investors claimed the financial risk was too great, but McKeithen kept the project on track with some shrewd political maneuvering: he convinced the state to lease the stadium from itself, becoming both landlord and tenant. By the time the hearings and lawsuits were settled, architects and engineers were at work tackling the problem of building the biggest domed structure on earth.

Construction began on August 11, 1971, and was spearheaded by Nathaniel "Buster" Curtis, a local architect who brought in specialists in the new field of computerized engineering. One of the first problems—how to support the tremendous weight of the building—was solved by a recent technological breakthrough that made it possible to insert prestressed concrete pilings into the soft soil. The stadium and parking lot would sit on 7,000 such piers, each 10 inches in diameter and extending to a depth of 175 feet.

The next challenge was designing the columns and braces that would support the roof and give stability to the structure. Resting on the columns would be a tension ring equipped with rocker bearings, allowing the roof to "float" with the expansion and contraction caused by changing outside temperatures. The dome's walls would essentially hang

from the roof, with force vectors at the foundation level pointing away from the center of the building. The walls would be wrapped in a skin of anodized aluminum. On the drafting table the plans looked fine, but even the designers were unsure whether the structure would hold the 5,000-ton roof.

Until June 12, 1973, the roof rested on jacks atop 37 steel towers and other temporary scaffolding. That day, the jacks were removed, one by one, and the dome settled onto the tension ring. Under its own tremendous weight, the building sank three and one half inches into its foundation—but the

be staged simultaneously. A wedding banquet or antique show can proceed quietly just yards from the noisy throngs at an athletic event.

In its first 20 years the Superdome hosted more Super Bowl games than any other venue. College basketball championships have also been played here. Among the musical stars who have performed under the dome are Frank Sinatra, Roy Orbison, and Aretha Franklin. The stadium was the place where Muhammad Ali regained his heavyweight title in 1978. The Republican Party nominated George Bush as its presidential candidate here in 1988. A year later, an even more important personage was the center of attraction: Pope John Paul II addressed 80,000 schoolchildren in the dome.

ECONOMIC
BOOM

In the 1960's, New Orleans, like many other cities across America, was in serious physical decline, with worn-out neighborhoods being abandoned and commercial zones begging for renewal and reclamation. The run-down area around Poydras Street in the

SLEEK AND EFFICIENT LINES
The graceful curve of the stadium's walls, above, complements the architecture of the neighboring buildings. A system of rain gutters rings the roof's perimeter. The gutters' 345,000-gallon capacity allows rainwater to be fed into the stadium's drainage system, thereby avoiding problems of flooding.

columns and the tension ring held. The largest steel dome in the world was solidly in place, literally holding the rest of the building together.

There was also the question of how to keep the gigantic structure and dome stable. The dome is a lamella configuration, a series of overlapping triangles that extend outward from a central ring at its apex. This configuration provided stability against downward pressure, but a counterbalance was necessary to protect it against upward forces. Providing the counterbalance is a 75-ton gondola, suspended from the center inside the stadium. Giant television screens, located on six sides of the gondola, enable spectators to watch action replays.

Dixon's vision of the Superdome as a multipurpose facility was accomplished in several ways. Numerous sections of arena seats were built on motorized dollies in order to tailor seating for individual events. For football games, the arena can accommodate more than 70,000 spectators; for trade shows and exhibits, it can hold 103,000 people; for smaller affairs, perhaps a concert for 14,000, the seating can be organized into a more intimate arrangement. A system of curtains and backdrops can efficiently bring about a total transformation.

The ice rink used one day for a figure-skating show can be converted overnight into an earthen terrain for motorcycle racing. A day later, the stadium can be set up for a music festival or a trade show. All events rely on a computerized lighting system that controls more than 27,800 lamps.

Adding to the Superdome's versatility are four soundproof ballrooms where separate events can

central business district was an eyesore of ailing warehouses and freight yards crisscrossed by railroad tracks, power lines, and rutted avenues.

But when the Superdome was built, 52 acres of dilapidated real estate were cleared for the stadium and parking. Since it opened in 1975, the economic impact on the city has been enormous. The Superdome has generated some $4.6 billion in revenue over the past two decades. Hotels, shopping plazas and office towers went up on adjacent land, and studies have concluded that the stadium has been directly responsible for $2 billion worth of new construction. In 1975 there were about 10,000 hotel rooms in the New Orleans area. By 1997 there were some 25,000, and most of them were within a mile of the Superdome.

For all its economic importance to the city of New Orleans, the Superdome stands primarily as a monument to boundless human ambition. Once deemed an impossible dream, this awe-inspiring man-made structure will be remembered as a tribute to the ingenuity and perseverance of the talented individuals who found a way to get it built.

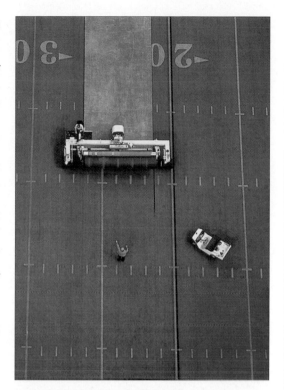

VERSATILE SURFACE
Workers remove one of the 26 panels of artificial turf used for football games, left. When the Superdome is used for football, more than 81,000 square feet of turf—nicknamed Mardi Grass—is laid down. Baseball requires more than 127,000 square feet. The stadium's concrete floor forms the base on which all other surfaces are laid, including wood flooring for basketball, ice for skating, and stages for concerts and other performances.

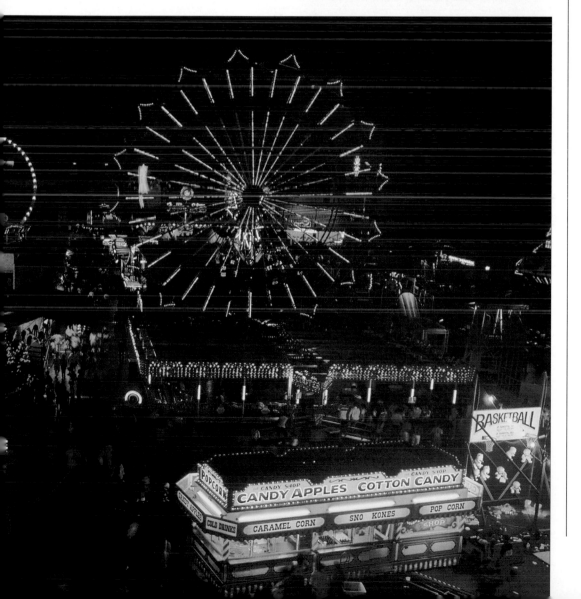

FUN FOR THE WHOLE FAMILY
Each May the interior of the Superdome is magically transformed into the Super Fair, left. An old-fashioned carnival, complete with cotton candy and a Ferris wheel, this annual event is just one of the many family-oriented activities staged here.

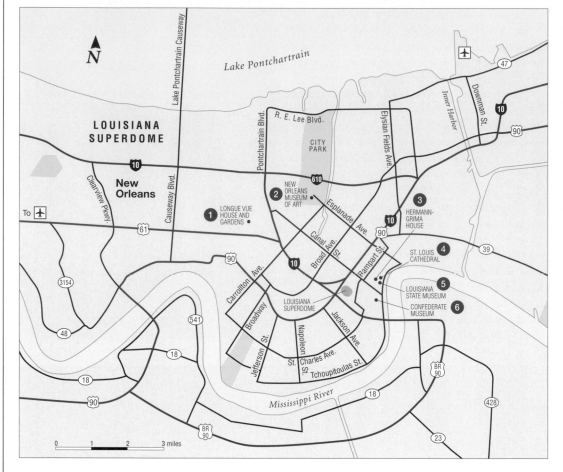

Longue Vue House, above, and its spectacular gardens were built by cotton broker Edgar Stern and his wife, Edith Rosenwald, the daughter of Sears, Roebuck & Co. magnate Julius Rosenwald.

① LONGUE VUE HOUSE AND GARDENS

This historic city estate was built between 1939 and 1942 by philanthropist Edgar Bloom Stern and his wife, Edith Rosenwald. The Classical Revival mansion is furnished with English and American antiques and needlework, and the floors are covered with Oriental and French carpets. The estate's collection of British creamware pottery is considered one of the best in the country. The mansion is surrounded by eight acres of formal gardens of magnolias, roses, azaleas, sweet olives, and live oaks. The Wild Garden features a trail that winds through a forest. Other gardens include the Pond, the Canal, and the rose-filled Walled gardens. Water fountains and mosaic sidewalks are part of a formal court found within the Spanish Court—Longue Vue's largest garden. Its design was inspired by the 14th-century gardens at the Alhambra palace in Granada, Spain. Famed American landscape designer Ellen Biddle Shipman designed both the gardens and the interior of the house. Located at 7 Bamboo Rd.

② NEW ORLEANS MUSEUM OF ART

This museum was founded in 1911 and showcases art from Africa, Europe, and Asia, as well as North, Central, and South America. Some 40,000 pieces in the collection are displayed in 26 permanent galleries. The Asian Collection showcases Chinese ceramics and bronze sculptures and 250 Japanese prints from the Edo period (1600–1868). A decorative arts exhibit focuses on glassmaking, from its origins in Egypt to the present; with more than 6,000 pieces on display, the collection of glass is one of the largest in the United States. This section also holds an extensive collection of Paul Storr silverware and features two period rooms: an 1810 Louisiana bedchamber and a 1790 Federal parlor. A large exhibit on European art from the 16th to the 20th centuries includes paintings by Edgar Degas, created between 1871 and 1872 when he was visiting relatives in New Orleans. Works by American artists include those of Georgia O'Keeffe, Jackson Pollock, and Jacob Lawrence. Visitors to the section of the museum devoted to Native American art can see kachina dolls from the Hopi and Zuni Pueblo peoples, Nez Perce beadwork, Pima and Apache baskets, and Pueblo pottery. Also on view are examples of pre-Columbian art, including Aztec and Mayan jade and gold jewelry. Located in City Park.

③ HERMANN-GRIMA HOUSE

This National Historic Landmark was built in 1831 by Samuel Hermann, an immigrant from Radelhein, Germany. The Federal-style mansion, a rare example of this type of architecture in New Orleans, features cypress doors, Corinthian columns, marble mantels, crystal chandeliers, and a large central hallway—an

in 1791 to function as the residence for the Capuchin monks of St. Louis Cathedral. The building now holds artifacts and exhibits on Louisiana culture and heritage, with displays focusing on photographs, Colonial-era paintings, and historic maps. A third structure, the Old U.S. Mint, was erected in 1835 and is the only building in the country to have served as the mint for both the U.S. government and the Confederacy. A small display recounts the history of the building, but the museum's main focus is on the history of New Orleans jazz. A large exhibit devoted to New Orleans' own Louis Armstrong displays his first trumpet, a cornet, and a custom-made gold trumpet mouthpiece. Also here are instruments owned by clarinetist Pete Fountain, drummer Gene Krupa, and pianist Bix Beiderbecke, along with an archive of photographs and 7,000 jazz recordings. The fourth historic building administered by the Louisiana State Museum is the 1850 House, which preserves one of two row houses built by Baroness Micaela Almonester de Pontalba. An apartment on the second floor is decorated with furniture and works of art typically found in a middle-class, ante-bellum New Orleans home. Located on Jackson Square in the French Quarter.

6 CONFEDERATE MUSEUM

The red sandstone Memorial Hall building, Louisiana's oldest museum, opened in 1891 as a meeting place for Confederate veterans who wanted to reminisce about the Civil War. A large Romanesque Revival–style building, it was designed by New Orleans architect Thomas Sully. An exhibition area with a cypress-wood ceiling and walls displays uniforms, guns, shells, mess kits, flags, Civil War–era medical instruments, and Gen. Robert E. Lee's field silver service. The museum also displays the personal effects of Confederate officers Braxton Bragg and P.G.T. Beauregard, as well as photographs and paintings of Civil War events that took place in Louisiana. Located at 929 Camp St.

The money to build St. Louis Cathedral, left, was donated by a wealthy Spaniard, Don Andres Almonester y Roxas, who stipulated that the congregation pray for his soul after his death. He died just four years after the cathedral was completed, and his body is buried beneath the church's marble floor.

An ornate wrought-iron balcony, below, graces the Upper Pontalba Building in New Orleans' historic French Quarter.

unusual feature in New Orleans interior design. The house is furnished with pieces from the mansion's subsequent owner, local judge and notary Felix Grima. Located at 820 St. Louis St.

4 ST. LOUIS CATHEDRAL

This triple-steepled church facing Jackson Square is the oldest active cathedral in the nation. It was built in 1794 on the site of two earlier churches, both of which met with disaster: the first was destroyed by a hurricane in 1722 and the other burned to the ground in 1788. Between 1849 and 1851 changes were made to the building: steeples were built onto the central and twin towers, and a classical facade was added. The interior contains frescoes, Renaissance-style murals, Latin inscriptions, and a three-manual Moller pipe organ built in 1950. In 1964 Pope Paul VI elevated the status of the cathedral to a Minor Basilica. Located on Chartres St.

5 LOUISIANA STATE MUSEUM

This museum administers four National Historic Landmark buildings found in New Orleans' famous French Quarter. One, the Cabildo, was constructed next to St. Louis Cathedral between 1795 and 1799 and served as the seat of the city's Spanish Colonial government. It contains displays of Colonial and military artifacts. The nearby Presbytère was built

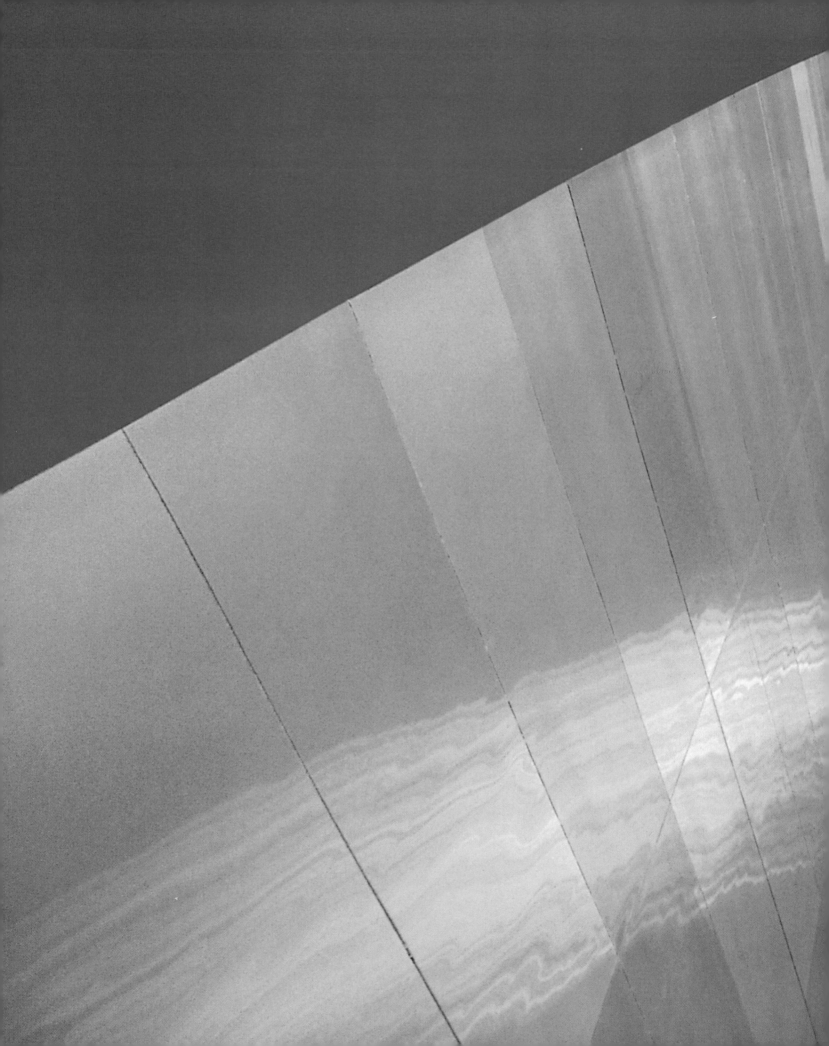

GATEWAY ARCH

*This St. Louis landmark traces
a graceful arc in the sky above
the city's riverfront.*

The idea to build the Gateway Arch was born in November 1933 aboard a railway passenger car chugging toward St. Louis. An attorney named Luther Ely Smith, returning home from a visit to the recently built George Rogers Clark Memorial in Vincennes, Indiana, was pondering the vital role St. Louis had played in the 19th century as the gateway to the Western frontier—and lamenting the sorry state it was in now. The city's glory days seemed long gone. The St. Louis waterfront, once a vibrant place familiar to countless pioneers and profiteers for its energy and promise, was now a blighted, ramshackle slum. Something was needed to rekindle the boom days of old St. Louis. At the sight of St. Louis on that bleak November day, Smith had a brainstorm: what the city needed was a monument, something on the scale of Vincennes' new classical white rotunda—a symbol that would celebrate St. Louis' history. A few days later Smith shared his idea with Bernard "Barney" F. Dickmann, the recently elected mayor of St. Louis.

Dickmann reacted with excitement. Why not erect a monument to the pioneer movement that settled the American West? After all, what era in American history was more deserving of memorializing? And what location was more appropriate than St. Louis—the Gateway to the West?

GATEWAY TO THE FRONTIER

St. Louis has a distinctive history. The city's origin dates to 1764, when French fur trapper Pierre Laclède set up a trading post on the west bank of the Mississippi. The post was soon surrounded by a village named for the patron saint of France—King Louis IX. St. Louis prospered, even as it was transferred from France to Spain and back to France, then sold by Napoleon Bonaparte to the United States as part of the Louisiana Purchase in 1803. President Thomas Jefferson bought the 828,000 square miles for $15 million, realizing that the nation-sized acquisition would enable the United States to achieve a coast-to-coast dominion. In the years that followed, St. Louis became the jumping-off point for explorers and adventurers, trappers and traders, freighters, farmers, pioneers and promoters, all eager to head west. Charged with protecting the steady stream of dreamers and doers were the troops of the U.S.

Army. The city's population also began to grow as more and more people stayed behind to offer their wares to those who were lured westward.

Amid the gloom of the Great Depression, the idea of constructing an expensive memorial must have seemed like a distant dream—mere political puffery. But Barney Dickmann was no political hack. He was determined to see the plan to completion. A former World War I gunnery sergeant, Dickmann was the son of German immigrants who had arrived in America in 1875. The veteran politician had worked hard to achieve his position as a prominent businessman. As a two-term mayor, Dickmann replaced slum dwellings with decent housing, expanded the city's health care, promoted commercial development, wrangled the construction of a memorial to the city's military veterans, and mobilized a coalition of leading citizens to find a way to free St. Louis from one of the country's worst air pollution problems.

Dickmann promoted his vision for a westward expansion memorial with the grit and gumption of an Old West pioneer. In April of 1934 he assembled a brain trust of St. Louis' leaders and named the group the Jefferson National Expansion Memorial Association; appropriately, Luther Ely Smith was elected as its chairman. Pushed and

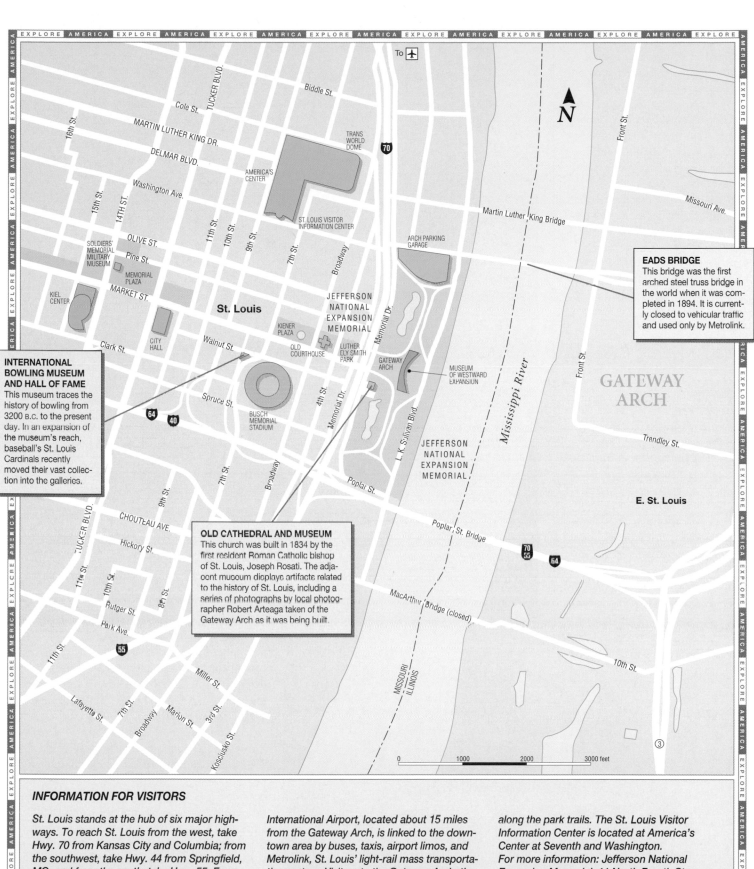

To ✈

N

TUCKER BLVD.

Biddle St.

Cole St.

16th St.

MARTIN LUTHER KING DR.

DELMAR BLVD.

Washington Ave.

15th St.

14TH ST.

TRANS WORLD DOME

70

AMERICA'S CENTER

OLIVE ST.

SOLDIERS' MEMORIAL MILITARY MUSEUM

Pine St.

MEMORIAL PLAZA

11th St.

10th St.

9th St.

7th St.

Broadway

St. Louis Visitor Information Center

Martin Luther King Bridge

Front St.

Missouri Ave.

ARCH PARKING GARAGE

KIEL CENTER

MARKET ST.

St. Louis

CITY HALL

Clark St.

Walnut St.

KIENER PLAZA

OLD COURTHOUSE

JEFFERSON NATIONAL EXPANSION MEMORIAL

LUTHER ELY SMITH PARK

Memorial Dr.

GATEWAY ARCH

MUSEUM OF WESTWARD EXPANSION

L. K. Sullivan Blvd.

Mississippi River

EADS BRIDGE
This bridge was the first arched steel truss bridge in the world when it was completed in 1894. It is currently closed to vehicular traffic and used only by Metrolink.

Front St.

GATEWAY ARCH

Trendley St.

INTERNATIONAL BOWLING MUSEUM AND HALL OF FAME
This museum traces the history of bowling from 3200 B.C. to the present day. In an expansion of the museum's reach, baseball's St. Louis Cardinals recently moved their vast collection into the galleries.

64

40

Spruce St.

BUSCH MEMORIAL STADIUM

4th St.

Memorial Dr.

JEFFERSON NATIONAL EXPANSION MEMORIAL

E. St. Louis

9th St.

7th St.

Broadway

Poplar St.

Poplar St. Bridge

70 55

64

TUCKER BLVD.

CHOUTEAU AVE.

Hickory St.

11th St.

10th St.

8th St.

OLD CATHEDRAL AND MUSEUM
This church was built in 1834 by the first resident Roman Catholic bishop of St. Louis, Joseph Rosati. The adjacent museum displays artifacts related to the history of St. Louis, including a series of photographs by local photographer Robert Arteaga taken of the Gateway Arch as it was being built.

MacArthur Bridge (closed)

Rutger St.

Park Ave.

55

11th St.

MISSOURI
ILLINOIS

10th St.

③

Miller St.

Lafayette St.

7th St.

Broadway

Marion St.

3rd St.

Kosciusko St.

0   1000   2000   3000 feet

INFORMATION FOR VISITORS

St. Louis stands at the hub of six major highways. To reach St. Louis from the west, take Hwy. 70 from Kansas City and Columbia; from the southwest, take Hwy. 44 from Springfield, MO; and from the south, take Hwy. 55. From Springfield, IL, in the north, take Hwy. 55 south; from the east, take Hwy. 70; from the southeast, take Hwy. 64. The Lambert–St. Louis International Airport, located about 15 miles from the Gateway Arch, is linked to the downtown area by buses, taxis, airport limos, and Metrolink, St. Louis' light-rail mass transportation system. Visitors to the Gateway Arch, the Museum of Westward Expansion, and the Old Courthouse can park in the garage on Washington Ave. and then walk to the arch along the park trails. The St. Louis Visitor Information Center is located at America's Center at Seventh and Washington. For more information: Jefferson National Expansion Memorial, 11 North Fourth St., St. Louis, MO 63102; 314-425-6010.

EXPLORE AMERICA EXPLORE AMERICA EXPLORE AMERICA EXPLORE AMERICA EXPLORE AMERICA EXPLORE AMERICA EXPLORE AMERICA EXPLORE AMERICA EXPLORE

GATEWAY ARCH 69

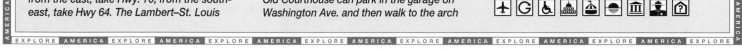

TIMELESS TRIBUTE
Architect Eero Saarinen chose the classic catenary arch, with its timeless beauty, as a fitting form for the memorial, right, to Thomas Jefferson and the countless pioneers who took the American dream westward. He wrote: "An absolutely simple shape—such as the Egyptian pyramids or obelisks—seemed to be the basis of the great memorials that have kept their significance and dignity across time."

PIECE BY PIECE
Saarinen's design called for an arch built of triangular sections, below, of stainless steel on the outside and carbon steel on the inside. The sections, which were built in Pennsylvania and shipped to the site, were hoisted into position by derricks—giant ground derricks for the first 72 feet, then smaller creeper derricks that were able to climb up the legs of the arch. The Gateway Arch took four years to erect and used 17,246 tons of building materials—including 5,119 tons of steel.

promoted by Mayor Dickmann, the group petitioned the U.S. Congress for matching funds. Federally funded New Deal ventures were unfolding across the land. Why not another one? With surprising speed, the association won a pivotal victory: on December 21, 1935, Pres. Franklin D. Roosevelt signed a bill into law that would fund the proposed memorial with a matching grant of $6.75 million. The citizens of St. Louis responded by approving a bond referendum, and within four years the National Park Service had acquired almost 40 acres of prime waterfront property for the site of the memorial.

The memorial campaign stalled temporarily during World War II. In 1947, however, the Memorial Association reignited the campaign by sponsoring a $225,000 competition for the memorial's design. The competition elicited 172 entries from architects and designers across the country.

mistakenly directed to his father, who had also entered the competition. Both family and firm had already celebrated with a champagne victory party for the father when they realized that it was the son who was the competition finalist.

UNUSUAL DESIGN Eero Saarinen's entry was the unanimous choice of the jury, who praised its arch-shaped design. "The arch," explained the architect, "could be a triumphal arch for our age as the triumphal arches of classical antiquity were for theirs." Saarinen's arch was not designed to be a flat-topped stone arch; instead he chose a catenary arch, which is the shape assumed by a chain when each end is held from two separate points. Furthermore, the Gateway Arch is a weighted catenary arch, which means that its curve bows outward as though weights are hung at strategic locations along the curve. It was to be made of stainless steel and would use a triangular structure to support its massive weight.

Almost immediately, Saarinen's design provoked controversy. The Italian dictator Benito Mussolini, someone recalled, had promoted a similar arch-shaped symbol for a wartime exposition in Rome. The comparison troubled some observers in post-war America, but Saarinen and his St. Louis sponsors held firm. Other critics denounced the proposed monument as a "a stupendous wicket" and "a big hairpin." Still others declared a city baseball stadium would have been a more practical monument. "I've just seen this steel monstrosity in St. Louis," declared Pennsylvania congressman Joseph P. Vigorita, who was to oppose additional federal funding of the memorial in 1966. "I was appalled. Why don't we just take an old battleship and give it to them to stand on end?"

Many critics, however, embraced Saarinen's arch and affixed almost mystical symbolism to the design by declaring it representative of mankind's never-ending search for new frontiers of progress. The shape, said one observer, "draws heavily upon the cosmic symbolism of the dome of heaven, an archetype that appears almost universally in the ancient world." The architect was more phlegmatic, admitting that he had been influenced by the Jefferson Memorial in Washington, D.C. "In a way, it's the same as our Jefferson Memorial in St. Louis—in one case the dome, in the other case the rounded arch," explained Saarinen. "I was thinking of the problem in that way, and only later did it occur to me that it was a gateway to the West."

Before the construction of the arch could get under way, there were many issues to be resolved. One of them threatened to derail the project:

It was 37-year-old Finnish-born architect Eero Saarinen who won the contract. Saarinen's father, Eliel, was also an architect, and his mother, Loja Gesellius, was a sculptor and artisan. Eero Saarinen was 13 when his family moved to the United States. He attended public schools in Michigan, where his father designed the Cranbrook Academy of Art. In 1929 Eero went to Paris, France, and studied sculpture. But, as he said years later, "it never occurred to me to do anything but follow in my father's footsteps." Two years later he entered the architecture program at Yale University. After graduating with honors, Eero joined his father's architectural practice, developing a modernistic style that utilized up-to-date technology. For some years Eero remained in his father's shadow. When he was named one of the finalists in the competition to design the Jefferson National Expansion Memorial in St. Louis, the congratulatory announcement was

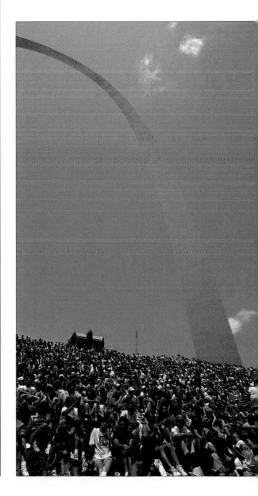

FOURTH OF JULY
Crowds of people gather beneath the Gateway Arch, below, to celebrate Fair St. Louis, an annual event, which is held during the Fourth of July weekend. Festivities include a parade, an air show, fireworks, and musical performances.

Heralded as the largest arch in the world, the Gateway Arch, above, was designed to be heavier at the ends in order to make it look even taller than it actually is.

A photograph taken in 1965, right, shows two cranes mounted on creeper derricks as the arch nears completion. Observing the arch at an earlier stage of construction, a journalist wrote: "To your right and left, workmen swarm around the foundations of the two legs. High overhead, the creeping derricks haul more materials into place. The legs now are high enough that your eye can trace a graceful line across the sky that will be the path of the completed Arch . . . . You'll never be able to discard the memory of one of the most spectacular and most difficult construction jobs in history."

riverside railroad tracks needed to be relocated to make way for the arch and the 91-acre park that would accommodate it. For years the project languished. Then, in 1957, St. Louis mayor Raymond R. Tucker suggested a workable new route for the tracks, and modifications to the design were implemented. Years of delays ensued, during which bids were called for and the site prepared. Construction finally began in 1962, and on February 12, 1963, the first stainless steel section of the arch was laid—atop a 45-foot-deep foundation set on bedrock.

Heralded as a "victory for engineering," the construction of the Gateway Arch skin required a staggering 886 tons of stainless steel—at the time, the largest amount of stainless steel ever ordered. The contract to build the massive arch was awarded to the MacDonald Construction Company, which had built missile silos for the U.S. government. The steelwork was subcontracted to the Pittsburgh–Des Moines Steel Company.

Each of the steel sections arrived from Pittsburgh on railroad flatcars. Both legs of the arch were built simultaneously by stacking the huge triangular-shaped sections on top of one another. Much of the work was carried out by gargantuan 100-ton creeper derricks mounted on special tracks built atop the legs of the arch. These and smaller-sized creeper derricks were manufactured specially to build the hollow, triangular-shaped arch, which had to accommodate an interior tramway and be sturdy enough to withstand tornado-force winds.

Slowly, the modern-day colossus inched its way up into the sky alongside the Mississippi. The residents of St. Louis craned their necks to study its daily progress, hoping that the towering contraption would not topple into the river. "Dimensions are so critical," observed the Gateway Arch's project manager, Ken Kolkmeier, "that we even do our surveying at night when the temperature on all three steel walls is equalized." Kolkmeier, who

TOP OF THE WORLD
Inside the towering apex of the arch, visitors, left, peer out of some of the 32 windows to take in the panoramic view. Colorful flags, below, set up for Fair St. Louis, flap in the wind above Memorial Drive.

at 30 years of age was even younger than Saarinen, understood the challenges of his job: an uncorrected error on the ground would prevent the two legs of the arch from connecting at the top.

Projections called for completion of the arch in 1963, but unexpected delays continued right up to the eve of the 1965 topping-off ceremony. Only two days before the October 28 festivities, the local ironworkers union called a work stoppage for a safety check. But after a National Park Service official pronounced that "the arch is perfectly safe," the work was resumed. Before television crews and a rapt audience of thousands, a huge crane hoisted the 10-ton keystone into place, while water, pumped from fire hoses that were hauled midway up the south leg of the arch, cooled the structure in order to ensure a proper fit.

DREAM COME TRUE

Smith and Dickmann's dream has been realized in the 630-foot arch that forms the centerpiece of Jefferson National Expansion Memorial, a 91-acre national park on the St. Louis riverfront. On a clear day the Gateway Arch can be seen 30 miles away. Its two massive legs, straddling a distance just over the length of two football fields, support a structure that is 325 feet higher than the Statue of Liberty and 75 feet higher than the Washington Monument.

More than 65 million visitors have toured the memorial since its completion. The park surrounding the base of the arch hosts celebrations and activities throughout the year. The adjacent Old Courthouse, believed to have influenced the design of the U.S. Capitol, was where the famous Dred Scott case began in 1846; the building has been preserved by the National Park Service as part of the memorial. An exhibition hall, the Museum of Westward Expansion, located in a cavernous chamber beneath the Gateway Arch, contains 20 interpretive displays focusing on the importance of St. Louis' role in settling the West. The museum exhibits some 200 artifacts related to St. Louis' past, as well as giant black-and-white and color photomurals inscribed with evocative quotations.

Tickets for the tramway to the observation deck at the top of the arch are available at the underground visitor center. The tramway, designed by engineering consultant Richard B. Bowser, operates on the principle of a Ferris wheel, moving eight barrel-shaped passenger capsules up one leg of the arch to the sky-high observation room and down the other. The trip to the top takes less than four minutes. From here, visitors can see St. Louis bustling with activity below, river traffic prowling the muddy Mississippi, and the Missouri and Illinois landscapes sprawling toward the horizon.

In the end it is the Gateway Arch—not the historic events and people it memorializes—that is the greatest attraction at Jefferson National Expansion Memorial. Such irony would not have surprised Eero Saarinen, who predicted that his great arch would stand for a thousand years. Saarinen himself never saw the completed arch standing tall beside the Mississippi. He died a year before its construction began, at the age of 51, during an operation to remove a brain tumor.

One person who did survive the 32-year campaign to build the arch was its premier champion: Barney Dickmann. The promoter lived to see the Gateway Arch restore the fortunes of downtown St. Louis and reinvigorate its economy. At the 1965 topping-off ceremony, Dickmann was characteristically extravagant. "This is the greatest memorial since the Eiffel Tower," he happily proclaimed. "I'm glad the darned thing's finished."

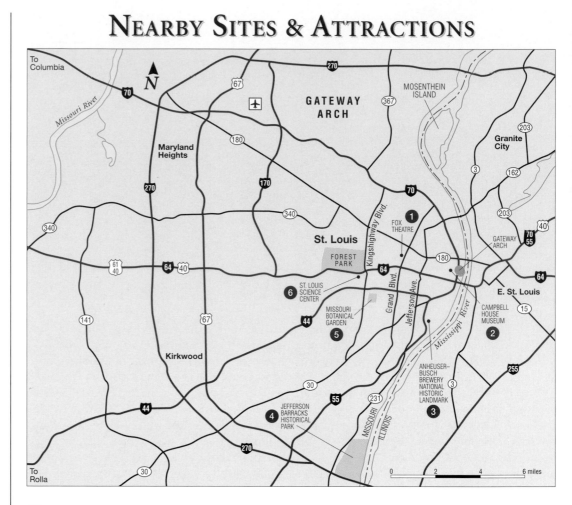

A laborers' house, below, is one of several restored buildings at Jefferson Barracks Historical Park. Americans who were stationed here include Ulysses S. Grant, Jefferson Davis, and Robert E. Lee.

1 FOX THEATRE

Opened in 1929 at a cost of $6 million, this Art Deco theater was one of the first in the nation to feature "talkie" films. The elaborate Siamese-Byzantine design includes jewel-encrusted columns, walls and ceilings decorated with plasterwork depicting exotic animals, and a 1-ton, 12-foot-in-diameter chandelier made of 2,264 pieces of glass. The theater houses a 4-keyboard, 2,700-pipe Wurlitzer organ, one of only five ever made. Restoration of the theater began in 1981: its 4,500 seats were restored and reupholstered, and 7,300 yards of carpet were woven with the theater's unique elephant pattern. Located at 527 North Grand Blvd. in St. Louis.

2 CAMPBELL HOUSE MUSEUM

This museum preserves the 1851 house owned by St. Louis businessman Robert Campbell. After working as a fur trader for the Rocky Mountain Fur Company, the Irish-born Campbell settled in St. Louis in 1835 and became a property owner and the president of two banks. The house's Rococo Revival style is apparent in the lavish design of the formal drawing room, morning room, and Victorian bedrooms. Beside the mansion is a large garden with a gazebo and a carriage house, which displays several family carriages. When the last of Campbell's three bachelor sons died in 1938, it was discovered that very few family items or furnishings had ever been discarded. In 1942 the house was turned into a museum to showcase the extraordinary collection of 19th- and early 20th-century household goods and furnishings that once belonged to the Campbell family. Located at 1508 Locust St. in St. Louis.

③ ANHEUSER–BUSCH BREWERY NATIONAL HISTORIC LANDMARK

Buildings on the tour of the world's largest brewery include the six-story 1892 Brew House, the Packaging Plant, Administration Building, and a horse stable. Copper kettles, ornate wrought-iron railings, and hop-vine chandeliers are on display in the Brew House, where a video presentation recounts the history of the company and its brewing process. Visitors can watch high-speed bottling and canning production in the Packaging Plant and see Clydesdale horses in the 1885 octagonal stable. The Anheuser empire began in 1860 when Eberhard Anheuser, the owner of a small brewery, teamed up with his brewery supplier son-in-law, Adolphus Busch. Located at 721 Pestalozzi St. in St. Louis.

④ JEFFERSON BARRACKS HISTORICAL PARK

This 1,700-acre barrack, which sits atop a bluff overlooking the Mississippi River, was constructed in 1826. The fort became a vital supply center for troops and munitions moving west of the Mississippi River. In the 1850's the stronghold served as the arsenal for the city of St. Louis; during the Civil War the army's largest hospital was set up here. The fort was decommissioned in 1946. Four buildings from the original structure have been restored. The 1857 powder magazine chronicles the history of the fort with displays of maps, uniforms, flags, photographs, and weaponry. Also located in the park, the Jefferson National Cemetery is the final resting place for more than 70,000 soldiers. Located 10 miles south of St. Louis off Hwy. 55.

⑤ MISSOURI BOTANICAL GARDEN

A rose and herb garden, English woodland garden and a Japanese garden are among the attractions found on this 79-acre site. Established in 1859 by Henry Shaw, this botanical garden is one of the oldest in the country. Shaw's 1851 Tower Grove House displays Victorian furniture, utensils, and clothing, and the 1882 Linnean House is the oldest public greenhouse in continuous operation in the U.S. Located at 4344 Shaw Blvd. in St. Louis.

⑥ ST. LOUIS SCIENCE CENTER

This science center encompasses more than 232,000 square feet in three buildings. The center contains more than 650 permanent exhibits on space sciences, technology, and ecology and the environment. Exhibits simulate earthquakes and tornadoes. Other highlights include a planetarium and the Exploradome, which houses traveling exhibitions. Located at 5050 Oakland Ave. in St. Louis.

The elaborate Victorian facade of the Administration Building, above, is one of several 19th-century buildings on view during a tour of the Anheuser-Busch Brewery.

The Taikobashi Bridge, below, spans one of the many ponds at the Missouri Botanical Garden. Established in 1858, the garden is one of the leading botanical research institutions in the world.

GLENWOOD CANYON DRIVE

A collaboration between engineers and environmentalists, this section of Interstate 70 has garnered international acclaim for its beauty and design.

Some people think all interstate highways should be like this 12-mile stretch of Interstate 70 through western Colorado's Glenwood Canyon. Bisecting some of the most scenic landscape in the country, Glenwood Canyon Highway combines environmentally sensitive highway design with innovative engineering and construction techniques. The road offers frequent opportunities to pull over and have a picnic, go for a walk, or take a bike ride. Travelers may catch a glimpse of bighorn sheep scampering along the rocky slopes or watch a kayaker perform pirouettes on the Colorado River.

From the project's inception, its planners realized that this highway would have to be unlike any other in the country. The challenge was to place a four-lane interstate route within the confines of a very narrow canyon that was hemmed in by 2,000-foot-high cliffs and coursed by a powerful, churning river—while preserving the sublime beauty of the region.

Twenty years of ingenious planning and design and 12 years of meticulous craftsmanship—1 year for each mile—went into its creation. In some places the east- and westbound lanes are terraced along the north side of the canyon. In other places the road crosses over a series of bridges that sweep gracefully along the canyon walls. Sometimes the roadway disappears into tunnels bored through the Rockies. The highway includes 6 miles of bridges and viaducts, 3 tunnels, 60 million pounds of steel, and 400,000 cubic yards of concrete. All of this did not come cheap—before all was said and done, the price tag totaled nearly $41 million a mile.

Long before the advent of the automobile, Glenwood Canyon seemed impenetrable to travelers. In 1874 government surveyors reported that "the river is in

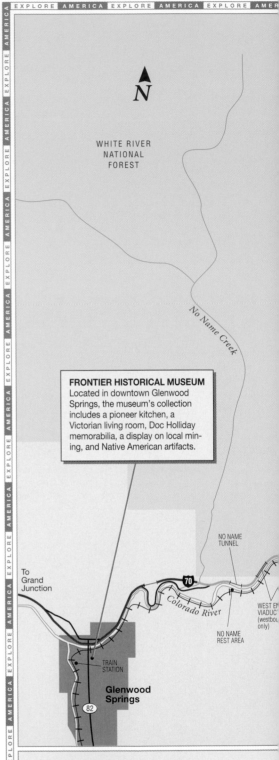

WHITE RIVER
NATIONAL
FOREST

No Name Creek

FRONTIER HISTORICAL MUSEUM
Located in downtown Glenwood
Springs, the museum's collection
includes a pioneer kitchen, a
Victorian living room, Doc Holliday
memorabilia, a display on local mining, and Native American artifacts.

NO NAME
TUNNEL

To
Grand
Junction

70

Colorado River

WEST EN
VIADUC
(westbou
only)

NO NAME
REST AREA

TRAIN
STATION

**Glenwood
Springs**

82

INFORMATION FOR VISITORS

*Glenwood Canyon Drive is located on a
12-mile stretch of I-70 between Glenwood
Springs and Dotsero. To reach the canyon,
take I-70 westbound from Denver for about
100 miles. The nearest airport is in Denver.
Because winter snowfall in the area can be
heavy, drivers are advised to use caution. In
the event of a breakdown, motorists should*

A HIGHWAY RUNS THROUGH IT
*With the waters of the Colorado
River rushing alongside them,
above, truckers transport their
goods along Interstate 70. The
highway was constructed using a
unique terraced design in which
the upper lanes slightly overlap the
lower ones. This was the engineering key to squeezing four lanes of
traffic through the narrow canyon.*

TUNNEL VISION
*Overleaf: Hanging Lake Tunnels
are one of three pairs of tunnels
dynamited through the rock for the
Glenwood Canyon Drive. State-of-
the-art technology within the tunnels includes carbon monoxide
detectors, fuel leak detectors, and
a radio rebroadcast system that
permits travelers to listen to their
radios while they are driving
through the tunnels.*

canyon. It is probably impassable to travel, the sides
being very steep. There is no Indian trail following the course of the river." When word got out in
the late 1870's and early 1880's that the canyon
might contain silver and coal, prospectors from
Aspen and Leadville came down side canyons in
order to penetrate the Grand River Cañon, as
Glenwood was then known.

In 1887 the Denver & Rio Grande Western
Railroad completed its line along the southern bank
of the Colorado River through Glenwood Canyon,
then shuttled silver and coal from mines in the surrounding mountains. Freight and passenger trains
still use the line, and in the 1980's lucky passengers could view the glories of the canyon and the
river from glass-domed train cars.

The first road to traverse Glenwood Canyon was
a simple two-rut, summer-only wagon road, which
was completed in 1902. Called the Taylor State
Road after Colorado senator Edward Taylor, this
path linked Denver on the eastern slope of the
Rockies with Grand Junction on the western slope.
According to a report by a local newspaper, resident
W. W. Price immediately drove an automobile
along the road, "endangering people's lives, by
frightening their horses." By the 1930's, however,
it was obvious that a larger highway was needed,
and the Works Progress Administration went to
work to transform Taylor State Road into the
paved, two-lane Highway 6/24.

In 1944 Congress authorized funds to initiate
the construction of the nation's interstate highway
system, but it was not until 1960 that money was
appropriated to extend I-70 westward from Denver.
Much of Highway 6/24 lies beneath an elevated

78

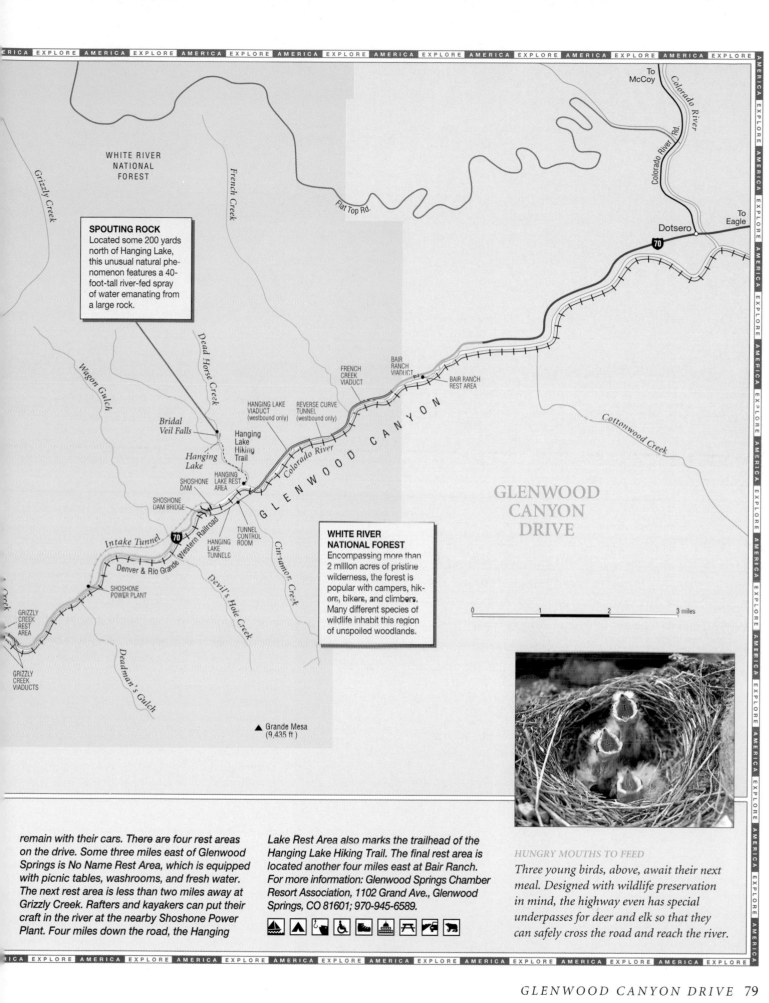

To McCoy

Colorado River

Colorado River Rd.

WHITE RIVER
NATIONAL
FOREST

French Creek

Flat Top Rd.

Grizzly Creek

To Eagle

Dotsero

70

SPOUTING ROCK
Located some 200 yards
north of Hanging Lake,
this unusual natural phe-
nomenon features a 40-
foot-tall river-fed spray
of water emanating from
a large rock.

Dead Horse Creek

Wagon Gulch

Bridal
Veil Falls

Hanging
Lake
Hiking
Trail

*Hanging
Lake*

SHOSHONE
DAM

HANGING
LAKE REST
AREA

SHOSHONE
DAM BRIDGE

HANGING LAKE
VIADUCT
(westbound only)

REVERSE CURVE
TUNNEL
(westbound only)

FRENCH
CREEK
VIADUCT

BAIR
RANCH
VIADUCT

BAIR RANCH
REST AREA

Cottonwood Creek

G L E N W O O D   C A N Y O N

Colorado River

Cinnamon Creek

Denver & Rio Grande Western Railroad

Intake Tunnel

70

HANGING
LAKE
TUNNELS

TUNNEL
CONTROL
ROOM

SHOSHONE
POWER PLANT

GRIZZLY
CREEK
REST AREA

GRIZZLY
CREEK
VIADUCTS

Devil's Hole Creek

Deadman's Gulch

**WHITE RIVER
NATIONAL FOREST**
Encompassing more than
2 million acres of pristine
wilderness, the forest is
popular with campers, hikers, bikers, and climbers.
Many different species of
wildlife inhabit this region
of unspoiled woodlands.

**GLENWOOD
CANYON
DRIVE**

0 · · · 1 · · · 2 · · · 3 miles

▲ Grande Mesa
(9,435 ft)

HUNGRY MOUTHS TO FEED

remain with their cars. There are four rest areas
on the drive. Some three miles east of Glenwood
Springs is No Name Rest Area, which is equipped
with picnic tables, washrooms, and fresh water.
The next rest area is less than two miles away at
Grizzly Creek. Rafters and kayakers can put their
craft in the river at the nearby Shoshone Power
Plant. Four miles down the road, the Hanging

Lake Rest Area also marks the trailhead of the
Hanging Lake Hiking Trail. The final rest is
located another four miles east at Bair Ranch.
For more information: Glenwood Springs Chamber
Resort Association, 1102 Grand Ave., Glenwood
Springs, CO 81601; 970-945-6589.

*Three young birds, above, await their next
meal. Designed with wildlife preservation
in mind, the highway even has special
underpasses for deer and elk so that they
can safely cross the road and reach the river.*

RICA EXPLORE AMERICA EXPLORE AMERICA EXPLORE AMERICA EXPLORE AMERICA EXPLORE AMERICA EXPLORE AMERICA EXPLORE AMERICA EXPLORE AMERICA EXPLORE AMERICA EXPLORE AMERICA EXPLORE

GLENWOOD CANYON DRIVE   79

Workers used an imported French gantry, above, to fit together concrete segments of the viaduct near Hanging Lake in 1990. The 350-foot-long device was employed in narrower stretches of the canyon where traditional equipment was too cumbersome. It is estimated that this single piece of equipment saved months of construction time and many thousands of dollars.

EXCITING EXCURSIONS

Rafters enjoy the gentle swells of the Colorado River, right. Several recreation businesses in the area offer rafting excursions through Glenwood Canyon.

section of Interstate 70; the rest was removed to make way for a recreation path that travels the length of the canyon.

The last remaining portion of I-70 through Glenwood Canyon proved to be the most challenging part to build. By the 1960's the nation's environmental consciousness was raised, and in 1969 the National Environmental Policy Act was passed into law. The act required that environmental impact statements be prepared for any major federal project. Clearly, such projects would include federally funded interstates. Concerned citizens' groups questioned whether a four-lane interstate could be built through the canyon without spoiling its beauty and serenity. If the project were to proceed, they demanded a design that ensured that traffic would travel at slower speeds. The groups also insisted that the canyon walls be left as undisturbed as possible, that the Colorado River be protected, and that only natural stone and native plants be used in landscaping along the highway. A citizens' advisory committee was appointed and was consulted in all stages of the planning.

In 1976 two alternate roadways were rejected and Glenwood Canyon was recommended as the route. A year later electronic surveying was under way and engineers produced maps, airbrushed photos, models, and even full-scale wooden bridge piers on-site to illustrate the highway's potential visual impact. Design proposals were subjected to endless rounds of hearings, reviews, media scrutiny, and a flurry of bureaucratic and legal maneuvers. Opponents of the project did not give up: at one point, the late John Denver, crooner of the popular song "Rocky Mountain High," threw a stone across the chasm to prove to the media that the canyon was too narrow for an interstate.

BUILDING A SHOWPIECE

Nevertheless, the project went ahead as planned, led by two inspired Americans who were both architects and engineers. Joseph Passonneau designed the western half of the highway, which called for separate terraced roadways. Edgardo Contini worked on the eastern end, which included some of the nearly 40 bridges and viaducts and the two tunnels built to avoid disturbing the trail at Hanging Lake. The plan was for the western and eastern work crews, totaling some 500 workers, to meet up in the middle.

Construction began in 1980 at No Name Creek. Without a detour around the project, as many as 18,000 vehicles a day had to be channeled into one lane and ushered through the canyon while heavy construction was under way. There were also some geologic surprises in store. The canyon walls were steep, unstable talus slopes of loose boulders. In order to stabilize them, 6,000 gallons of liquid grout were injected into the canyon walls. To the builders' amazement, the grout was simply swallowed up by the vast, empty spaces. Contractors returned to the drawing boards and settled on an experimental technique called compaction grouting, in which a stiffer grout was pumped into crevices at high pressure around bridge footings set into the slopes.

Rockfalls posed a potential hazard, so designers installed posts that were rigid enough to support fencing, yet flexible enough to spring back when falling rocks hit the fencing, thus causing the rocks to rebound back up the slope. Rock bolts and deflectors called rock "chimes" were placed in the gullies, and ice-shattering structures were installed below springs that typically froze in winter.

SNOWBOUND
A blanket of snow covers the Glenwood Springs train station, below. In winter this bustling community of some 8,000 offers skiing, ice climbing, snowboarding, and snowshoeing. Legendary gambler and gunslinger Doc Holliday succumbed to the ravages of tuberculosis here on November 8, 1887. He is buried in town.

KEEPING THE TRAFFIC ROLLING
Steady traffic streams through the Amphitheater Portals of the Hanging Lake Tunnels, left. More than $1 million of explosives were used to excavate the 4,000-foot-long tunnels. Today the tunnels are ventilated by eight fans, each capable of circulating 238,500 cubic feet of air per minute.

Terracing of the roadway required building retaining walls. To make the walls blend in with the rough surfaces of the canyon, vertical lines, or "rustications," were cut into the canyon's walls. The edge of the roadway was designed so that it jutted beyond the retaining walls, casting shadows that appear to lessen their height. To camouflage scars on the rock, the walls were sculpted and stained to match natural contours and colors.

Great lengths were gone to in order to preserve native plants. The French Creek viaduct threads between two 80-foot-tall Douglas firs that were saved. Another 150,000 trees and shrubs were replanted after construction. Willows, currant bushes, rabbitbrush, Gambel oaks, and native grasses flourish on the cliffs and beside the river.

The east- and westbound tunnels at Hanging Lake, located midway along the 12-mile stretch, presented the job's most daunting challenge. The two 4,000-foot-long tunnels took the highway over to the southern side of the river and kept the area of Hanging Lake free of cars. Some 250,000 cubic yards of rock had to be excavated, requiring the use of 150-ton cranes with equipment and workers hoisted up to the rock cuts.

Tunneling began with a 12-foot pilot bore to test the stability of the rock. Deeming it suitable, engineers had bolts fastened into the raw rock of the arched roof. On the strength of the report on the

The Glenwood Canyon project pioneered the use of posttensioned concrete slabs in the roadway itself. The technique, borrowed from airport runways, involved lacing a web of cables, putting the cables under immense tension, and then epoxying the cables into the concrete. Many of the bridges and walls were precast in sections off-site. The bridges were delivered by trucks in 40- to 55-ton concrete segments, which were then lifted into place by a 350-foot-high erection gantry.

rock's quality, local engineers convinced federal overseers that steel ribs were not needed to reinforce the tunnel walls—the first such tunnel in the United States. The Hanging Lake Tunnels were then lined with some 3 million gleaming porcelain tiles, each of them positioned by hand.

A four-story control room was constructed underground between the two tunnels. Electronic sensors constantly monitor tunnel traffic, weather changes, and emergency situations. With the opening of the Hanging Lake Tunnels in October 1992, the Glenwood Canyon portion of Interstate 70 was officially dedicated.

Four rest areas along the interstate—No Name, Grizzly Creek, Hanging Lake, and Bair Ranch—provide access to some of the canyon's natural riches. From the Hanging Lake Rest Area, a trail heads up the steep mountainside, where travelers are transported into a quiet world of deep green ferns and spruce trees. Mica-flecked boulders are splotched with silver lichens and festooned with thick moss. Red raspberries are there for the plucking. Hanging Lake, a cold aquamarine pool fed by a waterfall, awaits those who travel to the end of the trail. Hanging Lake Trail is strictly for those in good physical condition: the ascent is some 6,000

FUN ALONG THE ROAD

Today a bicycle path dissects the canyon, and bikes can be rented in Glenwood Springs and at the No Name Rest Area. Pedaling along the wide, paved paths between the eastbound lanes and the Colorado River, cyclists can sometimes spot a marmot perched on a rock or a belted kingfisher diving for a fish. Orange-jacketed paddlers in rafts churn through rapids, while more experienced boaters negotiate the stretch of white water between the Shoshone Power Plant and Grizzly Creek. The power plant and its diversion dam upstream have been harnessing the power of the Colorado River since 1909.

feet. Many who complete the hike record their accomplishment in the register at the trailhead. One well-earned boast from a hiker reads: "74 Year Old Grandma from Ohio Made It."

Anglers can try their skill at Grizzly Creek, where native whitefish spawn in late fall. And the warm, soothing waters of Glenwood Springs await both weary motorists and nature lovers alike.

Rarely has the marriage between technology and wilderness been so successful. While the execution of the Glenwood Canyon project was painstaking and intricate, the philosophy behind it was quite simple. In the words of state project manager Ralph Trapani, "We treated Mother Nature well."

NEARBY SITES & ATTRACTIONS

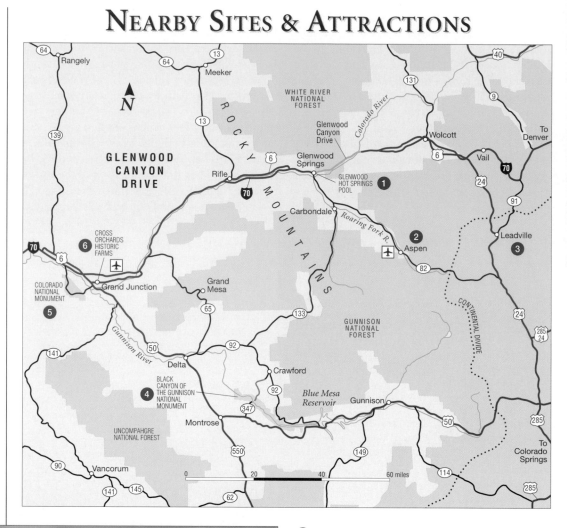

The panoramic scenery of the Rocky Mountains dominates the view of downtown Aspen, below. When the Ute Indians first discovered this region, they gave it a name that meant Shining Mountains. The resort's year-round population of 5,400 swells during the summer and winter months, when both outdoor enthusiasts and culture seekers arrive here to take advantage of Aspen's many attractions.

① GLENWOOD HOT SPRINGS POOL

Ute Indians were the first people to discover that the Glenwood hot springs could soothe their aches and pains. They called the springs Yampah, which means "Big Medicine" in their language. In 1888 developers hired the Viennese architect Theodore von Rosenberg to design a stone bathhouse for visitors. The building was completed in 1890 at a cost of $100,000. Today the hot springs consist of two pools: the therapy pool, which is kept at 104°F, and an enormous swimming pool with a temperature of 90°F; the latter has a diving area and lap lanes. Located on I-70.

② ASPEN

Magnificent scenery and some of the best skiing in the country have turned Aspen into an exclusive tourist resort and a retreat for millionaires and Hollywood celebrities. The town began as a couple of mining camps set up in 1879 after the discovery of silver in the Roaring Fork valley. By 1893 the town's annual silver production was $1 million a month and the population had increased to 14,000. Many of those who struck it rich built some of the grand homes on Hallam Street. During the height of the boom, wealthy Jerome B. Wheeler, an early citizen of Aspen, erected Hotel Jerome—still Aspen's

social hub—and the 1889 Renaissance Revival Wheeler Opera House, which stages first-rate musical, dramatic, and operatic productions throughout the year. Wheeler's residence, now known as the Wheeler–Stallard House Museum, is open to the public. Aspen's boom period did not last long. In 1893 the silver market crashed and most of the mines closed down. Then, in 1936, Ted Ryan and his partner, American bobsled racer Billy Fiske, saw the potential to transform Aspen and its surroundings into a ski resort similar to those in Europe. After they took the first steps toward developing the area, others followed suit. During the winter months the snow-covered slopes of nearby Aspen, Aspen Highlands, Buttermilk, and Snowmass mountains draw skiers from all over the world. Aspen's winter carnival, Winterskol, is held in January. From June through September, the town is the site of a variety of seminars and workshops, as well as musical and cultural festivals. Located off Hwy. 82.

3 LEADVILLE

Leadville burst onto the scene with the discovery of gold in 1860 in nearby California Gulch. The mines around the town, which was originally called Oro City, yielded millions of dollars worth of gold until they were played out. Then, in 1875, prospectors Alvinus Wood and William Stevens tested the gold mines' heavy sands and found that they contained nearly pure lead carbonate imbued with silver. When word got out, profiteers flocked to town, real estate boomed, and fortunes skyrocketed. One storekeeper named Horace Austin Warner Tabor grubstaked two prospectors. After they struck pay dirt on Fryer Hill, Tabor used his profits to buy a stake in other lodes, including the Matchless Mine. In 1893 he divorced his wife, Augusta, and married a blue-eyed young woman named Elizabeth McCourt Doe. But Tabor's rags-to-riches saga had an unhappy ending. In 1899, penniless and dying, he uttered his last words to his wife, known to history as Baby Doe: "Hang on to Matchless." She did, and when she died 36 years later, her body was found in the cabin above the mine. Today Tabor's picture hangs in the Hall of Fame in the National Mining Museum on 9th Street. The museum also displays mineral specimens and mining equipment, as well as walk-through replicas

of a hard rock mine and a coal mine. Tabor's Matchless Mine is now underwater, but the cabin in which Baby Doe died is on display. Mining artifacts and photographs from Leadville's boom days are on exhibit at the Heritage Museum and Gallery, which also presents works by contemporary Colorado artists. Located on Hwy. 24.

4 BLACK CANYON OF THE GUNNISON NATIONAL MONUMENT

This national monument covers 20,766 acres of mountains, forests, and canyons. Black Canyon itself cradles 12 miles of the deepest portion of the Gunnison River gorge; the width of the canyon measures about 1,100 feet at the top, but at river level it narrows in places to about 40 feet. Visitors can view the canyon from a dozen observation points, or overlooks, on the South Rim, reached by Hwy. 347 leading off Hwy. 50, 6 miles east of Montrose, or from the North Rim, accessible via an 11-mile gravel road from Hwy. 92 east of Crawford. The North Rim is closed in winter. Both rims have campsites and places to picnic. Hikers planning to descend into the canyon are required to consult a ranger at the visitor center at Gunnison Point on the South Rim. The center offers interpretive programs on the region's geological and natural history. The Gunnison Point Visitor Center is located 15 miles north of Montrose on Hwy. 347.

5 COLORADO NATIONAL MONUMENT

Juniper and piñon pines grow in this semidesert park that encompasses more than 20,454 acres of 500-foot canyons and plateaus. The Rim Rock Drive is a 23-mile road that skirts the edges of the canyons as it winds through the park, offering breathtaking views of the distant Colorado River valley, Book Cliffs, and flat-topped Grand Mesa. Visitors can see numerous unusual rock formations and isolated monoliths with such evocative names as Coke Ovens, Independence Monument, and Pipe Organ. Information on nature trails, hiking paths, camping and picnicking areas, as well as permits for back-country camping are available at the park visitor center near the Fruita entrance. Located west of Grand Junction off Hwy. 70.

6 CROSS ORCHARDS HISTORIC FARMS

Originally owned by the Cross family of Massachusetts, the 22,000 apple trees in this 243-acre orchard once yielded a long list of varieties—among them Black Twig, Gano, Jonathan, Winesap, Rome Beauty, and Ben Davis. From 1896 until 1923 the orchard was one of the largest in western Colorado. But an infestation of codling moths made apple production unprofitable, and the land was divided and sold at auction. Under the aegis of the Museum of Western Colorado, volunteers raised funds to buy back the land and buildings. Today displays interpret the social and agricultural heritage of western Colorado. Located on Patterson Rd. in Grand Junction.

The Tabor House, left, was built in 1877 by millionaire Horace Tabor. He lived here with his wife, Augusta, until 1881. Today the museum displays Victorian furniture, family portraits, toys, and photographic albums from the latter days of the 19th century.

The sheer walls of Black Canyon, below, are made of somber Precambrian rock, hence the canyon's name. Only experienced and well-equipped climbers are advised to tackle the rock face. Climbers are required to obtain a permit from park authorities before attempting any climbs.

HOOVER DAM

*Taming the Colorado River, the
Hoover Dam has transformed
a desert into a thriving region.*

At sunset the desert cools. Bats take wing and coyotes crouch amid the creosote bushes. A gusty breeze dances along the washes, drawing the night toward the sullen gloom of Colorado's Black Canyon. Suddenly, far below a hairpin bend in U.S. Highway 93, a soft but powerful light pours from the heart of the canyon, where the smooth arc of the Hoover Dam spans the rift of scorched rock. Structurally and esthetically, the dam is the epitome of grace under pressure. To those visionaries who built it, the dam's pearl-white crescent dramatically confirmed the conviction that human ingenuity could overcome any challenge that nature dared to throw in its path.

Perched on the Nevada wall of the canyon, the Hoover Dam's cylindrical visitor center provides tourists with an outstanding view of this marvel. Artifacts, films, and tours through the power plant, located 700 feet below the center, make for an unforgettable journey for visitors who learn the story behind the creation of this masterpiece.

From the Rocky Mountains to the Mexican border, the Colorado River drains some of the West's most breathtaking landscapes. Once,

PRODUCING POWER
The dam's generators, right, total 17 and occupy the powerhouse, which is located in 650-foot-long wings on either side of the canyon below the dam. Water reaches the turbines when four penstocks— two on either side of the Colorado River—are opened.

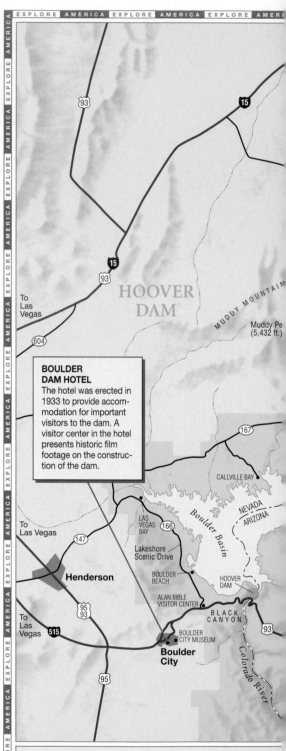

HOOVER
DAM

MUDDY MOUNTAIN

Muddy Pe
(5,432 ft.)

To
Las
Vegas

**BOULDER
DAM HOTEL**
The hotel was erected in 1933 to provide accommodation for important visitors to the dam. A visitor center in the hotel presents historic film footage on the construction of the dam.

CALLVILLE BAY

Boulder Basin

NEVADA
ARIZONA

To
Las Vegas

LAS
VEGAS
BAY

Lakeshore
Scenic Drive

BOULDER
BEACH

HOOVER
DAM

Henderson

ALAN BIBLE
VISITOR CENTER

BLACK
CANYON

To
Las
Vegas

BOULDER
CITY MUSEUM

**Boulder
City**

Colorado River

INFORMATION FOR VISITORS

Las Vegas, the closest city to the Hoover Dam, is located about 30 miles away. From the center of Las Vegas, take Hwy. 95 /93 for 20 miles to the junction with Hwy. 93. Travel 10 miles east on Hwy. 93 to the Hoover Dam. Kingman, AZ, is 68 miles from the dam via Hwy. 93. Major airports are located in Las Vegas and Kingman, which is served by the Mohave County Airport. The Hoover Dam, including the visitor center, is

the stormy river romped freely through alpine meadows, carved the sandstone badlands of the Grand Canyon in violent floods, and roamed at will across the barren low deserts of California. But the river's most daunting route was through Black Canyon, a jagged slot sliced into the 800-foot-high walls of volcanic andesite breccia.

As California developed in the early years of the 20th century, the river's unpredictable cycles of flooding and drought threatened newly prosperous towns and farms. Arthur Powell Davis, the second commissioner of the Reclamation Service (later called the Bureau of Reclamation), dreamed of harnessing and developing the Colorado through a system of water projects anchored by an enormous dam on the lower river below the Grand Canyon. In 1922 legislators from seven states, aided by then-Secretary of Commerce Herbert Hoover, agreed to a fair distribution of the river's water, thus clearing the path for a dam. From 1921 to 1923, surveyors investigated Boulder Canyon, but when they test-drilled 20 miles farther downstream in rugged Black Canyon, its thinner sediments, more cohesive bedrock, and larger reservoir capacity made Black Canyon a more attractive locale for the dam. Congress approved funds for the building of the dam in 1928. On December 21, 1928, President Hoover signed the act into law, setting in motion the ambitious project.

From 1929 to 1930, Bureau of Reclamation teams, led by Chief Engineer Raymond Walter and Chief Design Engineer Jack Savage, wrestled with

HARNESSING THE COLORADO RIVER
Overleaf: The downstream face of the Hoover Dam stands as a testament to the visionaries and hardworking men who erected the dam.

88

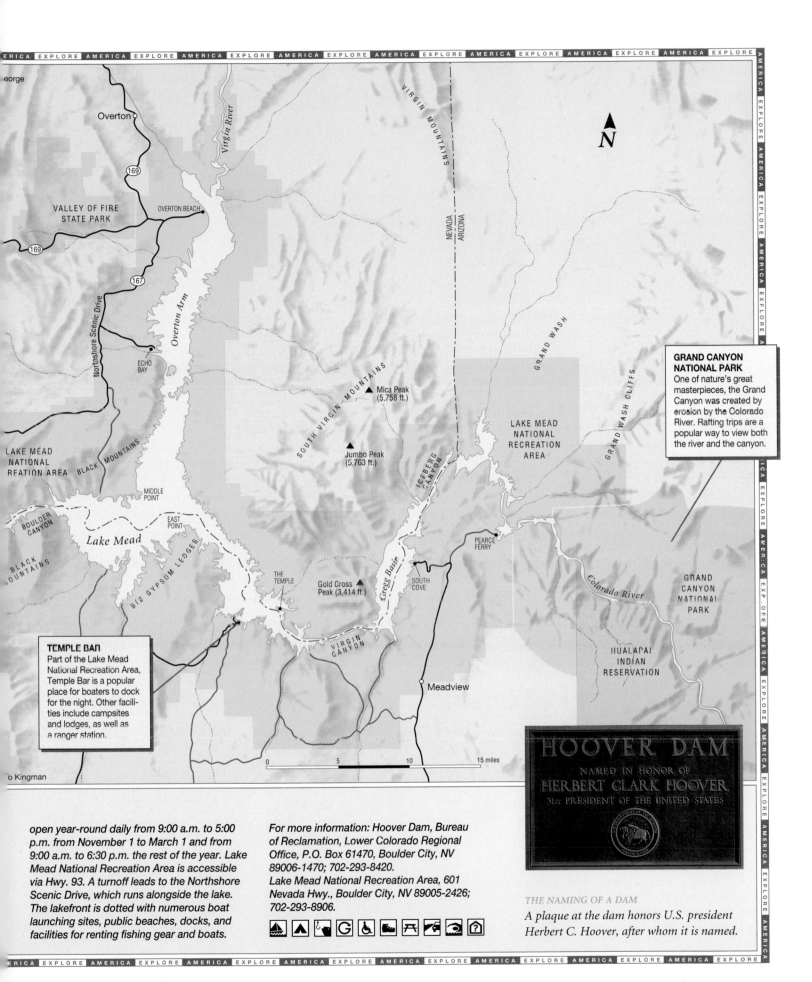

Overton

VALLEY OF FIRE
STATE PARK

Virgin River

Virgin River

169

169

167

Overton Beach

Overton Arm

Northshore Scenic Drive

ECHO
BAY

LAKE MEAD
NATIONAL
REATION AREA

BLACK MOUNTAINS

MIDDLE
POINT

BOULDER
CANYON

BLACK
MOUNTAINS

Lake Mead

EAST
POINT

BIG GYPSUM LEDGES

THE
TEMPLE

Gold Cross
Peak (3,414 ft.)

Gregg Basin

SOUTH VIRGIN MOUNTAINS

Mica Peak
(5,758 ft.)

Jumbo Peak
(5,763 ft.)

ICEBERG CANYON

VIRGIN
CANYON

SOUTH
COVE

PEARCE
FERRY

Meadview

VIRGIN MOUNTAINS

NEVADA
ARIZONA

GRAND WASH

GRAND WASH CLIFFS

LAKE MEAD
NATIONAL
RECREATION
AREA

Colorado River

GRAND
CANYON
NATIONAL
PARK

HUALAPAI
INDIAN
RESERVATION

o Kingman

0   5   10   15 miles

eorge

GRAND CANYON NATIONAL PARK
One of nature's great masterpieces, the Grand Canyon was created by erosion by the Colorado River. Rafting trips are a popular way to view both the river and the canyon.

TEMPLE BAR
Part of the Lake Mead National Recreation Area, Temple Bar is a popular place for boaters to dock for the night. Other facilities include campsites and lodges, as well as a ranger station.

HOOVER DAM
NAMED IN HONOR OF
HERBERT CLARK HOOVER
31st PRESIDENT OF THE UNITED STATES

open year-round daily from 9:00 a.m. to 5:00 p.m. from November 1 to March 1 and from 9:00 a.m. to 6:30 p.m. the rest of the year. Lake Mead National Recreation Area is accessible via Hwy. 93. A turnoff leads to the Northshore Scenic Drive, which runs alongside the lake. The lakefront is dotted with numerous boat launching sites, public beaches, docks, and facilities for renting fishing gear and boats.

For more information: Hoover Dam, Bureau of Reclamation, Lower Colorado Regional Office, P.O. Box 61470, Boulder City, NV 89006-1470; 702-293-8420.
Lake Mead National Recreation Area, 601 Nevada Hwy., Boulder City, NV 89005-2426; 702-293-8906.

THE NAMING OF A DAM
A plaque at the dam honors U.S. president Herbert C. Hoover, after whom it is named.

high as the tallest existing dam. At nearly a quarter of a mile wide and with a base thickness of 660 feet, it was also twice as large.

ART DECO
LOOK

Before the engineers could set to work on the dam, its final design needed to be completed on paper. From 36 different models submitted by different dam builders, a single design emerged: that of a graceful, arched wedge. Architect Gordon Kaufmann provided the final esthetic touches, removing the ornamentation from early plans and creating an Art Deco look. Kaufmann placed sleek observation towers along the rim to create a subtle play of light and shadow on the downstream face. He furnished the powerhouse with a simple facade of alternating columns and windows, and placed the transformers on the roof to accentuate the structure's purpose. He streamlined the spillways and topped the intake towers with cylindrical control houses. To heighten the drama, Kaufmann positioned beacons on the intake towers and flooded the canyon with a nighttime glow. Another architect, Allen True, designed the color schemes of the floors in the four towers, the walkways, and the power plant viewing platforms using Southwestern Indian

SIGHTSEEING ON THE LAKE

Lake Mead, above, created by the Hoover Dam, provides a perfect place for residents and visitors to beat the summer heat. Tour boats cruise the lake, which is popular for fishing and swimming. The nation's largest man-made body of water is named for Elwood Mead, dam sponsor and chief of the Bureau of Reclamation.

specifications for the gargantuan structure. They decided upon a massive concrete arch gravity dam. This design would not only resist the force of water by its own weight but also gain additional strength through arch action, which transfers the load to the walls of the canyon. Although gravity dams had already proven to be structurally sound, the Hoover Dam would have to hold back water pressure of 45,000 pounds per square foot. At 726.4 feet high, the Hoover Dam would be twice as

MULTIPURPOSE STRUCTURE

The Hoover Dam, right, serves a multitude of purposes. It irrigates more than 1 million acres of land in the United States and about a half million acres in Mexico; it provides water for the cities of Las Vegas, Los Angeles, San Diego, Phoenix, Tucson, and others; and it created Lake Mead, a popular recreation area that attracts about 9 million visitors each year.

motifs and polished black and green marble and terrazzo. Thirty-foot winged figures, cast in bronze, celebrate the people who planned and built the dam. More than a water project, the dam was designed to be a monument to the Machine Age.

When the design was finalized, the government called for bids. But many engineers doubted the dam could be built. They considered the site too remote and worried that the lake's immense weight might cause earthquakes. Not surprisingly, only three groups submitted bids. Of these, Six Companies, Inc., of Delaware stood out. This consortium represented an unlikely band of individualists who had built everything from roads, railways, and sewers to hotels, dams, and bridges. The men—Henry Kaiser, Felix Kahn, Warren Bechtel, Harry Morrison, Morris Knudsen, Charlie Shea—and their six firms had often been rivals. But they shared a risk-taking temperament, field experience in the West, and a passion for the dam. On March 11, 1931, Six Companies' low bid of $48,890,995 won the largest contract ever sponsored by the U.S. government. If they succeeded, engineering would be changed forever.

CAREERS IN THE MAKING

At the site of the dam itself, two men would play key roles. The first, the resident construction engineer, Walker Young, was a veteran of the 1920's surveys and the designer of the Arrowrock Dam in Idaho. He took charge of managing the project for the government. The nitty-gritty workings of the operation rested largely on the shoulders of Frank Crowe, general superintendent for Six Companies. Regarded as the best field engineer in the country, Crowe combined his ingenuity and efficiency with a hands-on style that quickly earned him the workers' respect. Crowe considered the Hoover Dam a "wonderful climax" to a brilliant career, but it was also a supreme personal and professional challenge.

Before construction could begin, Crowe had to tame the remote site. Black Canyon's sheer walls plunged directly into the river. Summer temperatures in the canyon soared above 120°F, while winter winds ripped through the gorge like runaway trains. Undaunted by nature's obstacles and the fact that there were no man-made amenities nearby, Crowe built an astounding mini-city called Boulder City. Some of the structures that went up included mechanized screening and mixing plants for the production of uniform concrete; rail lines to connect them to sand deposits and the dam site; roads in and out of the canyon; compression plants; blacksmiths' shops; and living quarters for thousands of workmen and their families. In addition, an endless march of transmission towers brought

power from a source 220 miles away. Unlike other hydroelectric stations, Hoover Dam's main structure would not contain the power plant. Instead, water would flow around the dam through 56-foot diameter, 4,000-foot-long tunnels in the canyon walls to a power plant at its base, or toe. In round-the-clock shifts, men on three-tiered trucks, called jumbos, attacked the rock with drills and dynamite. Muckers with pick and spade and, later, steam shovels loaded rock for removal into a steady stream of waiting trucks. Meanwhile, the canyon reverberated with spectacular explosions as high-scalers descended the precipitous cliffs of the canyon in rope harnesses to drill and blast loose rock from the walls, creating a clean abutment that would serve to anchor the dam. Working hundreds of feet above the canyon floor was exciting, prestigious, and high-paying, but dangerous. More than one high-scaler plunged to his death.

DANGEROUS WORK

A team of drillers attached to safety ropes, above, work on the canyon wall located above what became the location of the power plant. In the summer the walls of Black Canyon got so hot that they would literally scorch any bare skin that came in contact with them. During the day average temperatures often rose above 120°F and soared to 140°F in the tunnels. Of the 112 people who lost their lives during the project, 13 died of heat prostration.

The general view of Black Canyon, above, is taken looking upstream toward the site of the Hoover Dam. The photograph shows the downstream outlets of the diversion tunnels that curved the Colorado River around the dam site.

On November 14, 1932, trucks began dumping more than a million cubic yards of rock into the Colorado, diverting the raging river into the four tunnels that had already been bored through the canyon's walls: two on the Nevada side and another two on the Arizona side. With the river harnessed behind cofferdams—earth and rock structures used to seal off the construction areas—excavation began on the foundation. Steam shovels gouged the sediments of millennia until the bedrock lay exposed. The river had been tamed.

THE DAM RISES

The Boulder City Museum recounts the story of the people who made the dam—not just the high-profile engineers and politicians, but also the women and men who persevered in the dark days of the Great Depression. Machines made the Hoover Dam feasible, but the project was above all a human triumph. Prior to the establishment of Boulder City, single men lived in grim barracks in Black Canyon amid unrelenting heat and poor sanitation. Families and government employees occupied a squalid cluster of cardboard shanties known as Ragtown, where many died of dehydration beneath the desert sun. Workers toiled amid the danger of rockfalls in the canyon and the horrible possibility of carbon monoxide poisoning in the tunnels. Boulder City was a place where workers could get away and enjoy family life, but even this site was far from paradise. Construction engineer Walker Young described it as "greasewood and sagebrush, a few burros, tarantulas, centipedes, snakes." A company town that provided amenities under the watchful eye of government agents, it was also a "dry" town, and men seeking a wilder time trekked into Las Vegas. Today Boulder City is a community of parks and tree-lined streets, with a time-warp feel to its main street and sidewalk arcades. Now, as then, the magnetic pull is over the rise, toward the canyon.

When the time came to pour concrete, Frank Crowe's ingenuity shone. Refining his earlier innovations, Crowe strung a 1,200-foot cat's cradle of five fixed and traveling cableways across the canyon, enabling operators to pluck buckets filled with 16 tons of concrete from waiting railcars and deliver them with precision to their preordained forms. Pouring began on June 6, 1933, and continued without interruption for the next two years.

A CONCRETE STORY

Heat proved to be the biggest enemy of working with concrete: concrete shrinks with heat loss and swells with heat gain. When mixed with water, the oxides, silicates, and carbonates that make up cement undergo a chemical change: their crystals bond tightly to form strong concrete. A side effect of this hydration process is the release of energy as heat. Small dams diffuse heat through surface cooling, but the solution would need to be different at the Hoover Dam. Cooling the dam's 3.25 million cubic yards of concrete to air temperature could take up to 125 years. Added to this problem, the design had no margin for error—uneven cooling could cause the dam to crack, lose stability, and fail. A new low-temperature concrete mix proved to have some merit, but the real solution was structural.

The designers divided the dam into 215 interlocking, yet independent, columnar blocks, ranging from 25 to 60 feet square. Each block was filled with five feet of concrete at a time. One-inch pipes were placed in the concrete at each five-foot interval, and water, cooled to 38°F, was circulated through the pipes to help lower the temperature of the concrete. Some 582 miles of piping were used in the construction of the dam. A gigantic refrigeration plant cooled the water—300 gallons of which flowed through the piping system every minute. This ingenious system knocked about 120 years off the estimated time it would have taken the concrete to cure without the use of the cool water pipes. As one operator later remarked, the dam was just like "a huge radiator." Although it had been tested elsewhere, this cooling process received its first full-scale test at the Hoover Dam.

The work of pouring the concrete was a demanding task. The process relied on the skills of alert signalmen who acted as the eyes of cableway operators, who were often positioned out of sight of both railway and dam. Once the concrete was delivered, puddlers in high rubber boots stomped it into place. When the concrete was cured, the forms were moved up five feet, pipe fitters installed the refrigeration pipes, an adhesive grout was spread on top of the concrete so that the next five-foot layer would adhere to the one beneath it, and the next layer of concrete was poured. In this way, the interlocking sections of the dam rose within the canyon.

Today an elevator takes visitors down to the powerhouse. Here, 17 sleek generators hum away, fed by the river through an elaborate system of pipes and tunnels to produce more than 4 million kilowatt hours of electricity each year. The system is simple but massive. Each of the four larger penstocks that deliver water to the generators can carry 22,500 gallons per second.

GETTING THE FULL STORY
The new visitor center, above, completed in 1995, provides an interpretative program on the dam, including films and exhibits. Visitors on the tour can examine the hydroelectric generators and gaze at the dam's crest 560 feet above them as they stand on the power plant deck. From the overlook, the view of the dam, its powerhouses, Lake Mead, Black Canyon, and the Colorado River, 770 feet below, often elicit exclamations of awe from visitors.

A NATURAL TEMPLE
An outcropping called the Temple, left, in the Lake Mead National Recreation Area glows in an early evening sunset.

AN ELEGANT SPAN

The bird's-eye view at right shows the white arc of the Hoover Dam and the Colorado River as it twists through the rugged landscape. The river got its name from its sometimes red waters: it was named Rio Colorado, or "Red River," by the Spanish. The water runs red when it carries silt from the sedimentary rocks through which it has carved its course.

As water is discharged from the hydroelectric generator turbines, left, it flows into the Colorado River below the dam.

A 30-foot-high Art Deco bronze sculpture, below, is one of the two that stand beside the circular terrazzo floor of the Winged Figures of the Republic *monument. The work is a dramatic memorial to the dam's planners, designers, and builders.*

As the monumental dam neared completion, the attention of the nation focused upon Black Canyon. The dam was hailed as a triumph of American ingenuity, grand government thinking, and entrepreneurial know-how. For many people it was a final chapter in the winning of the West. On February 1, 1935, all but one of the 50-foot-diameter tunnels that had diverted the Colorado River were closed and Lake Mead was slowly born as the river once again flowed behind the massive wall of concrete. On September 30, 1935, Pres. Franklin Delano Roosevelt honored the tireless efforts of the planners, designers, engineers, and workers in his moving dedication, which was broadcast nationwide over the radio: "The transformation wrought here in these years is a 20th-century marvel." The first generator began to produce electric power on October 26, 1936, the second on November 14, and a third on December 28. As the demand for power increased, new generators were added. Today there is a total of 17 generators in the power plant.

The Hoover Dam's power has helped make Los Angeles a thriving metropolis. Its regulated water supply allows agriculture to bloom in the desert and has made Las Vegas one of the world's most popular resorts. And the lessons of its construction have made erecting high dams routine. The Hoover Dam is no longer the highest or biggest dam in the world, but in many ways it is still the best. To look up on its polished contours is to relive the drama of its birth. Echoes of Franklin Roosevelt's dedication, "I came, I saw, and I was conquered, as everyone would be who sees for the first time this great feat of mankind," still seem to reverberate amid the walls of Black Canyon.

NEARBY SITES & ATTRACTIONS

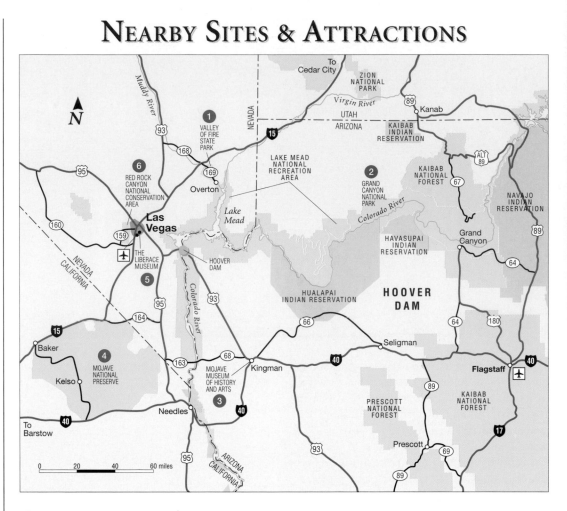

A Baldwin piano tiled with mirrors, below, stands center stage at the Liberace Museum. Eighteen of the flamboyant entertainer's pianos are on display at the museum.

1 VALLEY OF FIRE STATE PARK, NEVADA

A diverse landscape of brightly colored sandstone, limestone, and shale rock formations is preserved in Nevada's first state park. Located in a six-mile-long, four-mile-wide valley, the oddly sculpted terrain with its petrified sand dunes was created by millions of years of geological uplifting, faulting, and erosion.

Petroglyphs sketched onto the faces of jagged canyon walls, arches, and boulders trace the long human presence in the region. Archeologists date the petroglyphs to between 330 B.C. and A.D. 1150, when the Basket Maker people and Anasazi Pueblo farmers frequented the area. Plant life in the hot, dry valley is dominated by creosote, burro bushes, and numerous species of cacti. Although wildlife is difficult to spot during daylight hours, the park is home to snakes, lizards, and coyotes; birds in the area include the roadrunner and sage sparrow. There are two campgrounds, and a visitor center interprets the park's ecology, history, and geology. Located 55 miles northeast of Las Vegas on Hwys. 15 and 169.

2 GRAND CANYON NATIONAL PARK, ARIZONA

One of the world's most spectacular geological attractions, this national park preserves the monumental canyon carved by the Colorado River. The river has exposed layers of sediment, displaying 2 billion years of the earth's history. Nine hiking trails descend from the South Rim, some of them leading to the floor of the 277-mile-long canyon. Four trails on the North Rim descend into the canyon. Visitors can also view the canyon by horseback or by mule, or take a rafting trip along the river. Located 80 miles north of Flagstaff off Hwys. 180, 89, and 64.

③ MOJAVE MUSEUM OF HISTORY AND ARTS, ARIZONA

This museum uses artifacts, photographs, and artwork to chronicle the history of northwestern Arizona from prehistoric times to the present. Dioramas and murals in the front lobby depict the history of settlement in the area. A reconstructed wickiup brush shelter is displayed in the Hualapai Indian Room, along with a collection of kachina dolls, baskets, and pottery. An outdoor display exhibits farming and mining equipment and a 19th-century wooden caboose. Located at 400 West Beale St. in Kingman.

Ancient Anasazi granaries, above, are well hidden high above the Colorado River in Grand Canyon National Park.

④ MOJAVE NATIONAL PRESERVE, CALIFORNIA

This desert landscape of mountains, sand dunes, volcanic spires, cinder cones, and mesas was designated the first National Scenic Area in 1980. Hundreds of miles of gravel and dirt roads crisscross the 1.9-million-acre preserve, leading to campgrounds, ruins of old ranches and homesteads, mines, and rocks, which are inscribed with petroglyphs and pictographs estimated to be more than 10,000 years old. Near the town of Kelso lie the Kelso Dunes, where the slopes are so steep that the cascading sand creates a booming sound. More than 300 species of animals roam the preserve. Golden eagles and several species of hawks are drawn here by the warm thermal updrafts. The vegetation of Joshua trees, rabbitbrush, and creosote is typical of the desert. Located 63 miles east of Barstow on Hwys. 15 and 40.

⑤ THE LIBERACE MUSEUM, NEVADA

This museum pays tribute to Wisconsin-born pianist Liberace. Known by his many fans as Mr. Showmanship, Liberace had a penchant for the unusual. Eighteen of the 39 pianos that Liberace owned are on display in the museum, including a rhinestone-covered Baldwin, a mirror-embellished Baldwin, a 1788 Broadwood Grand, and a Pleyel piano that was played by Frederic Chopin. The Car Gallery contains Liberace's 1934 Mercedes Excalibur encrusted in Austrian rhinestones, his Rolls-Royce covered in mirror tiles, and his Volks Royce, a Volkswagen manufactured to resemble a Rolls-Royce. The Costume Gallery displays part of the glitzy entertainer's million-dollar fur collection, as well as his diamond-studded, piano-shaped ring. Located at 1775 East Tropicana Ave. in Las Vegas.

⑥ RED ROCK CANYON NATIONAL CONSERVATION AREA, NEVADA

This park preserves 196,000 acres of multicolored layers of limestone and sandstone hills and canyons. The geological timepiece was created millions of years ago when sand dunes, believed to have been blown here by the wind, hardened over a limestone bed. Over time, the sand dunes eroded into unusual shapes. Formed of red, orange, pink, yellow, and purple sandstone and older gray limestone, the sculpted hills are a spectacular setting for rock climbing. The cooler, steep-walled canyons are also popular among rock climbers and hikers. The park can be seen from a 13-mile loop drive, which climbs the lower rim of the 7,000-foot-high Red Rock Escarpment. A visitor center provides information on the geology, flora, and fauna of the area. Located 15 miles west of Las Vegas off Hwy. 159.

Petroglyph Canyon, left, bears the traces of human habitation in Valley of Fire State Park. The petroglyphs in the park are estimated to be more than 2,000 years old.

GOLDEN GATE BRIDGE

*Like an orange lyre, this glorious
bridge creates its own poetry as it
sweeps across the Golden Gate.*

Glowing in the first rays of day or silhouetted
against a setting sun, San Francisco's Golden
Gate Bridge seizes the imagination with its magnificent web of cables and struts. Commanding its
location with complete authority, the bridge's
aura of serenity belies the dramatic story behind
its conception, design, and construction.

Well into the second decade of the 20th century
San Francisco still relied on such ferry boats as
the enormous 1890 *Eureka,* which regularly carried up to 2,300 passengers and 120 cars across the
bay in a single trip before it was retired to the
local maritime museum. While many citizens
treasured the ferries' unhurried pace, others
believed that in order to fulfill its potential as a
world-class city, San Francisco needed a bridge
linking its peninsula to the mainland to the north.

One such visionary was City Engineer Michael
O'Shaughnessy, an indomitable Irishman who
had supervised the rebuilding of San Francisco
after the 1906 earthquake. O'Shaughnessy was
fond of asking visiting engineers how they would

CELEBRATING THE BRIDGE

On May 24, 1987, some 300,000 people (more than half of the population of the city proper) came from all over the country and crowded onto the bridge's six traffic lanes, right, to celebrate the Golden Gate Bridge's 50th anniversary.

BEAUTY UNSURPASSED

Overleaf: The entire length of the Golden Gate Bridge is 8,981 feet, but the main span between the towers is 4,200 feet long. Considered by many to be the most beautiful bridge in the world, the Golden Gate no longer boasts the world's longest bridge span. In 1964 the length of its main span was surpassed by the 4,260-foot-long Verrazano-Narrows Bridge in New York City; in 1981 a 4,626-foot-long bridge across England's Humber River claimed the record. Two longer bridge spans are scheduled to open in Denmark and Japan in 1997 and 1998, respectively.

100

go about bridging the Golden Gate strait. A good, hard look at the site invariably discouraged all but the most utopian dreamers. More than a mile wide, measuring over 300 feet deep at the center, swept by tidal currents ranging from 4.5 to 7.5 knots and by the outflow from seven major rivers, and alternately blasted by gales and shrouded in soupy fogs, the Golden Gate was no place to build a bridge. To make matters worse, the strait lay seven miles from the San Andreas Fault, which had unleashed the disastrous 1906 earthquake that leveled San Francisco. And, in any case, both shores of the strait were owned by the War Department—a government body that rejected the idea of a bridge, arguing that a single enemy bomb could destroy it and bottle up the entire harbor.

AN EXPENSIVE IMPOSSIBILITY In 1918 O'Shaughnessy sought to meet Joseph B. Strauss, a civil engineer known for his moveable bridges. "Everyone says it can't be done," Strauss later recalled O'Shaughnessy saying, "and even if it could be done, it would cost over $1 hundred million dollars." The crusty engineer rose to the challenge implicit in O'Shaughnessy's words and declared that he could build the bridge—and do it for much less money than the estimated price tag.

In his 1986 book The Gate, bridge historian John van der Zee ascribes a colorful list of attributes to Joseph Baermann Strauss, including "dreamer, tenacious hustler, publicity seeker, recluse." The

INFORMATION FOR VISITORS

Hwy. 101, which crosses the Golden Gate Bridge, runs between Washington's Olympic Peninsula and Los Angeles. Visitor services, including a gift center and a café (both presided over by a statue of chief engineer Joseph Strauss), are located on the southeast side of the bridge. To reach the area from the south, take the last San Francisco northbound exit off Hwy. 101, proceed right, then left into the east parking lot. From the north, travel along the far right toll lane #1, take the first right exit off Hwy. 101—the first right onto the road that passes underneath the Toll Plaza—and then proceed into the east parking lot. The area is part of the Golden Gate National Recreation Area, which encompasses a number of sites around San Francisco Bay, each offering a different perspective on the bridge. The view of the bridge from Alcatraz Island, once the site of the notorious prison, accentuates the immensity of the span. Pullouts along Conzelman Road in the Marin Headlands on the north side provide excellent views of the bridge. Walking trails in nearby Mt. Tamalpais State Park offer visitors many different vistas of the famous span.
For more information: The Golden Gate National Recreation Area, Fort Mason, Bldg. 201, San Francisco, CA 94123; 415-556-0560.

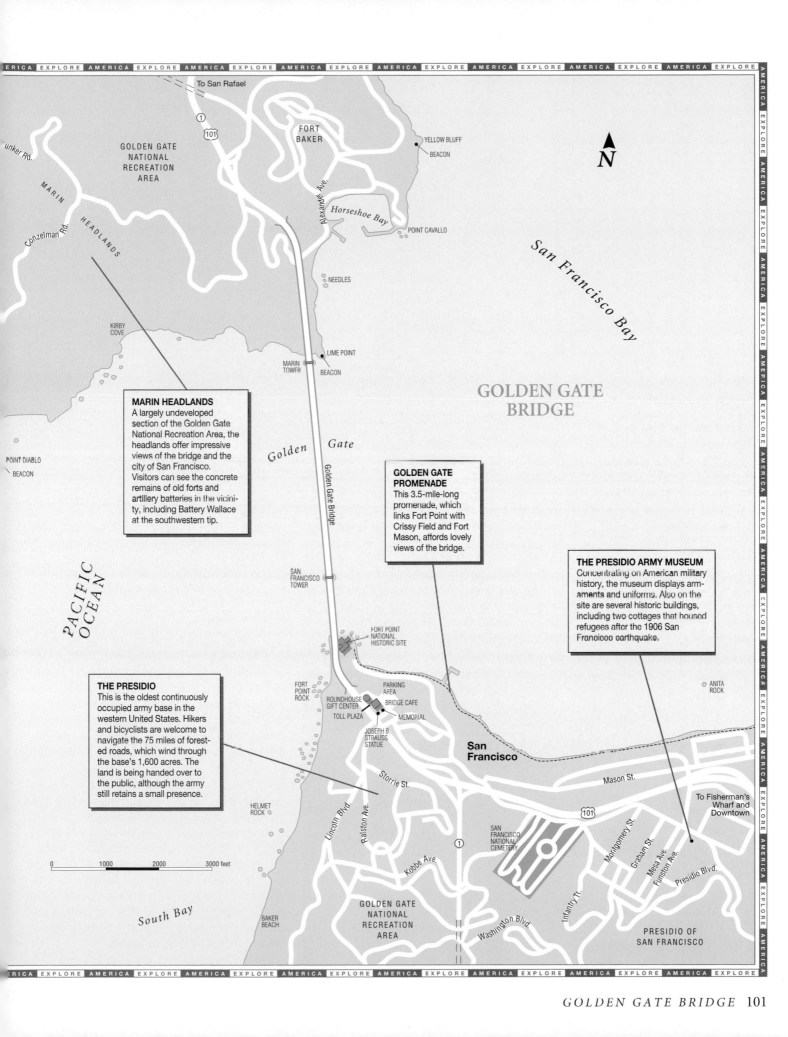

To San Rafael

FORT BAKER

YELLOW BLUFF
BEACON

GOLDEN GATE
NATIONAL
RECREATION
AREA

Horseshoe Bay

POINT CAVALLO

San Francisco Bay

N

NEEDLES

Golden   Gate

**GOLDEN GATE
BRIDGE**

KIRBY
COVE

LIME POINT

MARIN
TOWER   BEACON

Golden Gate Bridge

MARIN HEADLANDS
A largely undeveloped section of the Golden Gate National Recreation Area, the headlands offer impressive views of the bridge and the city of San Francisco. Visitors can see the concrete remains of old forts and artillery batteries in the vicinity, including Battery Wallace at the southwestern tip.

POINT DIABLO
BEACON

**GOLDEN GATE
PROMENADE**
This 3.5-mile-long promenade, which links Fort Point with Crissy Field and Fort Mason, affords lovely views of the bridge.

THE PRESIDIO ARMY MUSEUM
Concentrating on American military history, the museum displays armaments and uniforms. Also on the site are several historic buildings, including two cottages that housed refugees after the 1906 San Francisco earthquake.

**PACIFIC
OCEAN**

SAN
FRANCISCO
TOWER

FORT POINT
NATIONAL
HISTORIC SITE

ANITA
ROCK

THE PRESIDIO
This is the oldest continuously occupied army base in the western United States. Hikers and bicyclists are welcome to navigate the 75 miles of forested roads, which wind through the base's 1,600 acres. The land is being handed over to the public, although the army still retains a small presence.

FORT
POINT
ROCK

PARKING
AREA

ROUNDHOUSE
GIFT CENTER   BRIDGE CAFE

TOLL PLAZA   MEMORIAL

JOSEPH B
STRAUSS
STATUE

**San
Francisco**

HELMET
ROCK

Storrie St.

Mason St.

To Fisherman's
Wharf and
Downtown

101

Lincoln Blvd.

Ralston Ave.

SAN
FRANCISCO
NATIONAL
CEMETERY

Montgomery St.

Graham St.

Mesa Ave.

Funston Ave.

Presidio Blvd.

Kobbe Ave.

1

0   1000   2000   3000 feet

BAKER
BEACH

South Bay

GOLDEN GATE
NATIONAL
RECREATION
AREA

Washington Blvd.

Infantry Tr.

**PRESIDIO OF
SAN FRANCISCO**

ERICA EXPLORE AMERICA EXPLORE AMERICA EXPLORE AMERICA EXPLORE AMERICA EXPLORE AMERICA EXPLORE AMERICA EXPLORE AMERICA EXPLORE AMERICA EXPLORE

GOLDEN GATE BRIDGE 101

ambitious engineer had invented the trunnion bascule bridge, a drawbridge using concrete blocks as counterweights. Beginning in 1904, Strauss' firm built nearly 400 of these innovative bridges over rivers in the United States, Asia, Africa, South America, and Russia. It was the building of just such a small, useful structure that had brought him to the San Francisco area and O'Shaughnessy. Although he had no experience with large-span bridges, Strauss dreamed of leaving his mark and had the ambition to tackle the impossible.

Shortly before O'Shaughnessy issued his challenge to Strauss, the San Francisco Board of Supervisors had ordered a survey to determine whether a bridge could be built across the Golden Gate. In 1920 engineers took the first sounding off Fort Point and noted: "Federal experts believe it will be impossible to put piers at this point owing to strong currents and great depth."

FIRST ATTEMPT

While another man might have blanched at this evaluation, Joseph Strauss took up the gauntlet. In 1921 he presented O'Shaughnessy with a preliminary sketch of a bridge and an estimate of $25 to $30 million. His sketch was of an ungainly cantilever-suspension hybrid. Not surprisingly, the design fueled debates.

While ferrymen grumbled about losing their monopoly and esthetes decried the design's ugliness, Strauss barnstormed the northern counties for support. In this he was remarkably successful. Beginning in 1923, six of these counties joined San Francisco as the Golden Gate Bridge and Highway District, mandated to finance and build a bridge. The following year the War Department approved the plan. Still, lawsuits over the district's authority kept the project in limbo for another five years. It wasn't until 1929 that the district finally selected Joseph Strauss—a specialist in small bridges—to build the world's largest bridge.

Strauss immediately set about assembling a work force of 100 engineers and consultants—many of

pension bridges, the opposite is true: a thinner, more flexible truss transfers the load more effectively to the cables. A flexible truss was lighter, and lightness meant fewer materials and lower costs—a compelling combination for the board of engineers. San Francisco had found its bridge.

Having decided to erect a suspension bridge, the Board of Directors of the Golden Gate Bridge and Highway District still had to figure out how to over-

whom had once been his rivals to head up the contract. Among his expert advisers were Leon S. Moisseiff, who had pioneered the mathematics of suspension bridges; Othmar Ammann, who had supervised the design and the construction of New York City's George Washington Bridge; and Prof. Charles Derleth, chairman of the Department of Engineering at the University of California.

UNDER THE GUN

The bridge project's board of engineers convened for the first time in August 1929, under extreme pressure. Time and money were in short supply, and the district needed a blueprint before November, 1930, when voters would be asked to approve the bonds to finance the construction. There was a larger problem, too: Strauss' 1921 bridge was no longer economically feasible or technically viable. A new bridge was needed—and fast.

Fortunately, five years earlier Los Angeles architect Charles Rush had come up with a concept for a sleek suspension bridge that had impressed city engineers. The esthetics were acceptable and so was the science. Suspension bridges consist of a roadway hung from cables strung across the tops of two massive towers. Instead of withstanding the forces of nature with brute strength, they offset them with flexibility, balancing the framework of diagonal girders that carry the roadbed between the cables and a truss. In conventional steel bridges, stability is increased by making the truss more rigid. In sus-

come the obstacles that were inherent in building a roadway across the Golden Gate. No one had ever built towers as tall as those that would be needed. Nor had a bridge ever been built that had spanned such a wide harbor.

In the end, the challenges of erecting the bridge were reduced to mathematics—the calculation of stresses. The task fell to Charles Ellis, Strauss' chief structural engineer and a professor of structural and bridge engineering at the University of Illinois. Working against the clock, Ellis employed 33 algebraic equations using between 6 and 30 unknown quantities; these equations enabled him to work out all potential stresses—earthquakes, tides, wind deflection in the cables and roadway, and expansion and contraction of the cables in changing temperatures—to name but a few.

WORKABLE PLAN

With the numbers in hand, the designers could now proceed in concrete and steel. Two towers would rise 746 feet into the sky and support the cables that were held by immense concrete anchors at each end. The cables would support a truss 220 feet above the waters of the bay. The suspension design made possible a center span that was 4,200 feet long. This meant that the foundation for the northern, or Marin, tower would be on land, a bonus that resulted in great financial savings. The bridge's high clearance, with the towers supported by cross-bracing below the roadbed, allowed designers to replace cluttered overhead crossbeams with horizontal struts.

Strauss suggested a stepped-off design reminiscent of the Empire State Building's wedding cake tiers. After consulting John Eberson, an architect who specialized in theater design, he designed the towers so that they narrowed toward the top—a technique that emphasized their soaring height.

The towers' Art Deco flourishes are the work of Irving Morrow, a native of Oakland. Sensitive to the beauty of the bridge's physical setting, Morrow

MILITARY PAST

The Golden Gate Bridge crosses over Fort Point National Historic Site, above and right, at the bridge's southern approach. Built between 1853 and 1861, this four-story brick fortress was one of about 30 forts constructed as part of a national coastal defense system put into place between the War of 1812 and the end of the Civil War. In 1933, when the bridge was begun, the fort was still part of the Presidio, a working military reserve. Today interpreters lead tours through the batteries that were once armed with 126 cannons, none of which were ever fired in combat.

dams—temporary dams designed to keep water away while excavating to bedrock. Then the concrete foundations for the piers took shape. Work on the Marin pier progressed relatively smoothly, but the San Francisco pier required additional ingenuity; this pier was to be located 1,125 feet offshore. To get men and supplies out to the site, Strauss constructed a trestle that was strong enough to support heavy trucks and other equipment. A fender-ring, nicknamed the "giant bathtub" by workmen, was constructed at the end of the road. The ring enclosed the site of the pier.

On August 14, 1933, a freighter lost its way in heavy fog and demolished about 100 feet of the road. The road was repaired only to be washed

A visitor pauses to touch a cross-section of the bridge's main cable, left, outside the gift center on the San Francisco side. The accompanying sign itemizes the cable's vital statistics, including the fact that some 80,000 miles of wire were used to spin the main cable.

fashioned rectangular tower portals and placed bands of fluted surfaces on the struts to create a play of light and shadow. He also designed open railings, which allowed motorists to enjoy the view, and decided to paint them International Orange to blend with the bridge's natural setting.

AT LONG LAST Construction on the bridge officially began on January 5, 1933. Workers blasted into the slopes on both shores in order to place the anchorages. These huge blocks of concrete were embedded with eyebars that would secure the cables. While the anchorages progressed, work proceeded on the piers for the giant towers. To build them, crews needed to construct coffer-

away in a sudden storm in the fall of that year. Workers built a third, and stronger, road and hurriedly constructed a cofferdam made of 27-foot-long and 6-inch-thick concrete sections inside the fender-ring. Once the water was pumped out of the cofferdam, work could begin on the pier.

Meanwhile, the north tower was rising. Plates, girders, and diaphragms were boxed and assembled into groups of steel cells at workshops in Pennsylvania and shipped to the site, where the cells were riveted. In November 1934, workers crowned the Marin tower with two 150-ton cast steel cable saddles, which were designed to hold the cables in place. Working double shifts, laborers on the San Francisco tower set their saddles in place the following June.

PUTTING IT TOGETHER

An archival photograph, above, shows the roadway attached to suspender cables being assembled in sections. Conceived in the Roaring Twenties when Americans were optimistic about the future, the bridge was built during the Depression when millions were out of work. Competition to work on the bridge was fierce and the pay was low, ranging from $5.50 per day for unskilled workers to $11 per day for skilled iron workers.

The J. A. Roebling Company, founded by the man who designed the Brooklyn Bridge, was the obvious choice for the cable work because the firm had devised a method for spinning cables in midair. The cables would be the largest fabricated to date—about 3 feet in diameter and 7,650 feet long. Each cable consisted of 27,572 wires, grouped in 61 strands of 452 wires.

SPINNING THE CABLES On August 2, 1935, the Golden Gate was closed to shipping traffic so that a pulley cable could be strung from shore to shore. The cable was needed to help string up two large wire ropes across the strait, which were then attached to the concrete anchorages on each shore and slung over the top of the towers. Now the spinning plants at the two anchorages went into action, with their array of derricks, drive engines, and unreeling machines. Six spinning carriages, running back and forth on a tramway between the anchorages and a switching station at the center of the main span, played out the strands that eventually would become the main cables. The strands were gathered hexagonally and strung into their saddles atop the towers. Next, six compacting machines, consisting of a ring of hydraulic jacks,

rolled along the length of the cable on wooden spools, pressing the hexagon into a circle just over 36 inches in diameter. Wrapping machines encased the bundle of strands in a thin wire coat that immediately received a coat of International Orange paint. Cable bands, some weighing up to four tons, were clamped onto the main cables as anchors for the suspension ropes.

With the cables in place, work began on the roadway. To preserve stability, a lattice of riveted steel beams was extended in sections from the towers in both directions simultaneously. Like the tower cells, the individual units were prefabricated in shops and transported to the site, where they were moved into position by a traveling derrick and assembled by hand. As each segment was installed, crews adjusted the towers' saddle jacks in order to keep both of the towers vertical under the growing load of the roadway.

On November 18, 1936, the two center-span traveling derricks met, completing the truss, and workers prepared it to receive a concrete road surface. Crews simultaneously poured 20-by-50-foot sections of the road, working from each tower toward the middle. Small spacings allowed the steel truss to expand and contract with changing temperatures without damaging the roadway.

Thus far work on the bridge had proceeded with only one fatality, a credit to Strauss' extensive safety precautions. Among other measures, a safety net suspended under the roadway saved 19 lives. But on February 17, 1937, when clean-up crews were removing forms, a scaffold tore loose, ripped through the net, and plunged into the bay along with 12 workers. Ten of these men died, a tragic blemish on an otherwise remarkable safety record. Still, building the $27 million bridge produced fewer fatalities than expected, given that the norm on such a large-scale engineering project during the 1930's was one man killed for every million dollars spent. Shortly after the accident, work was resumed and the approaches and toll plaza received their finishing touches. The Golden Gate Bridge was officially open for business.

On May 27, 1937, more than 200,000 people crowded onto the bridge for Pedestrian Day, kicking off a week of celebrations. The next day cars roared across in procession, the first trickle in a stream that has not stopped since. In 1996, 41 million vehicles crossed the bridge.

A TRIUMPH OF GRACE

Joseph Strauss and his team had finished the job on time and under budget, stilling not only the furious beast of the Gate, but their critics. Today the bridge is an indelible part of San Francisco's identity. Ironically, one of the best ways to appreciate its elegant form is to catch a view of it aboard the Golden Gate-Sausalito ferry. As the sun sets, the water shimmers like iridescent mercury below the bridge's gossamer silhouette—a triumph of structural gracefulness and an engineering marvel that deserves its place as one of America's most beloved landmarks.

DWARFED BY A GIANT
A windsurfer challenges the Golden Gate's changeable winds, above. Gale-force winds have forced the closure of the bridge three times. Dwarfed by the supporting structure of the Golden Gate Bridge, the lighthouse, left, was built on the roof of Fort Point. The fort was decommissioned in 1933.

NEARBY SITES & ATTRACTIONS

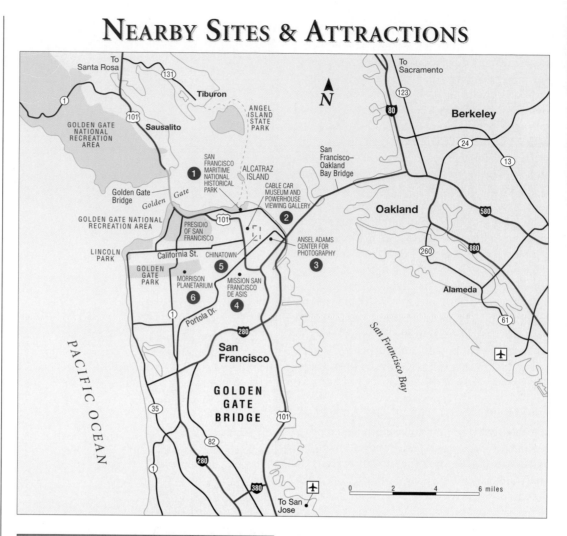

Mission San Francisco de Asis, right, known locally as Mission Dolores, is flanked by the basilica, which was rebuilt in 1918 after its destruction in the 1906 earthquake.

1 SAN FRANCISCO MARITIME NATIONAL HISTORICAL PARK

Located at the west end of San Francisco's bustling Fisherman's Wharf, this national historical park encompasses a maritime museum and seven historic vessels moored at Hyde Street Pier, as well as a swimming beach and a library with archival materials devoted to U.S. West Coast maritime history. The Maritime Museum's immense collection includes models of famous vessels, photographs, figureheads, and sections of boats that once plied the Pacific. The park's star attractions, however, are the vessels tied up at the Hyde Street Pier. Visitors can tour the *Balclutha,* a three-masted ship built in Scotland that once transported European coal to the United States; the steel-hulled vessel was featured in the 1935 film *Mutiny on the Bounty,* starring Clark Gable and Charles Laughton. Visitors can also tour the *Eureka,* a side-wheeler ferry with a four-story steam engine, and the 1902 steam tugboat from England's River Clyde. Located at the foot of Polk St.

2 CABLE CAR MUSEUM AND POWERHOUSE VIEWING GALLERY

In the 1870's, when former mine cable designer Andrew Hallidie saw how hard San Francisco's horse-drawn wagons had to struggle to climb Nob

Hill's steep streets, he was inspired to develop the world's first operational cable car. Hallidie's Folly, as early detractors referred to Hallidie's invention, involved towing trams along a thick cable that ran in a slot set between tracks and was driven by a steam engine in the powerhouse. Today San Francisco is home to a museum that is a veritable lesson in the cable car. An underground gallery displays the massive driving wheels, or sheaves, that keep modern cable cars running. Upstairs, visitors can examine vintage photographs and sections of some of the city's earliest cable cars, including Hallidie's 1873 prototype. A film demonstrates how the system works. Located at 1201 Mason St.

3 ANSEL ADAMS CENTER FOR PHOTOGRAPHY

Once a talented music student, native San Franciscan Ansel Adams began to shift his interest from music to the visual arts in the 1920's. At that time he discovered the satisfaction of experimenting with different styles of photography. Over subsequent years, Adams pioneered numerous innovative techniques for taking photographs of sweeping panoramas. This museum houses a gallery devoted to Ansel Adams' work. Other galleries display rotating exhibits of works by contemporary photographers, as well as exhibits that examine the history of photography. Located at 250 4th St.

4 MISSION SAN FRANCISCO DE ASIS

Also known as Mission Dolores, this mission contains an 18th-century chapel and an early 20th-century church. The whitewashed chapel is all that remains of the old mission that was completed by Franciscan monks in 1791. The sixth of 21 Spanish missions founded by the Franciscan order in California under the direction of Father Junípero Serra, Dolores was first established in 1776 about four blocks south of its present location. Poor soil prompted the monks to relocate the mission to its present site in 1782. The only intact mission in the original mission chain, the chapel is also the oldest intact building in San Francisco. A look at the ceiling, with its rough-hewn roof timbers tied together with rawhide, shows visitors how the chapel was built.

The ceiling is decorated with Ohlone designs that were applied when the chapel was restored. The parish church was completed in 1918 after the 1906 earthquake destroyed the original parish church. It was designated a basilica (an honorary church of the pope) in 1952 by Pope Pius XII. It includes design elements from the Corinthian and Moorish styles of architecture. At the rear of the church is a beautiful stained-glass window that depicts Saint Francis of Assisi, patron of the mission and the City of San Francisco. The mission cemetery contains a statue of Father Junípero Serra as sculpted by Arthur Putnam, an early California artist. Located on the corner of 16th and Dolores streets.

5 CHINATOWN

Crammed into a 16-square-block area on the eastern slope of Nob Hill in San Francisco, this vibrant center of the West Coast Chinese community is home to an estimated 30,000 people. The area was first settled in the 1840's by Chinese immigrants looking to escape the poverty and starvation in their homeland. Today visitors stroll the neighborhood's crowded streets and shop for exotic foods and imported goods from Asia. The Chinese Historical Society of America, located on Commercial Street, recounts the history of the Chinese in this country with period photographs and a fascinating collection of artifacts. Bounded by Stockton, Kearny, Bush, and Broadway streets.

6 MORRISON PLANETARIUM

Guided tours are offered at this planetarium, called the region's "largest indoor universe." A 5,000-pound Star Projector regularly screens spectacular shows that display more than 3,800 stars, moons, and planets on the planetarium's 65-foot-diameter domed roof. The planetarium offers star shows daily throughout the year, but more frequently during the summertime. The shows reveal how the universe was formed; they also chronicle the history of astronomy and of 20th-century exploration of the galaxy. The lasers used in the laser shows are hand operated and are choreographed to an eclectic musical score. Located within the Golden Gate Park.

The Balclutha, one of the vessels on display at San Francisco's Hyde Street Pier, bears a replica of the original demi-woman figurehead, left, that was bolted to the boat's bow in the Glasgow dockyards in 1886. The 12-foot-high replica statue was carved in 1987 in honor of the ship's centennial.

In San Francisco's Chinatown, street signs, like the one shown below, are written in Mandarin, as well as in English.

MONTEREY BAY AQUARIUM

*Space-age wizardry immerses
visitors in an undersea world—
without ever getting them wet.*

Behind the world's largest window, viewers are mesmerized by the sight of a soupfin shark gliding silently through shafts of light. Only a few feet away, schooling yellowfin tuna cruise past a giant ocean sunfish, and bulky sea turtles swim by with astonishing grace. Each year almost 2 million people visit Monterey Bay's revolutionary aquarium, where architects and engineers, spurred by a dedicated group of sea lovers and dreamers, have created a wondrous world by masterfully blending nature and artifice.

Monterey seems created for tourists. Rustic adobe buildings herald the city's past glory as the capital of Spanish California, while nearby Big Sur and Carmel offer rugged scenery and polished works of art. The area's popularity is not a recent phenomenon. When the first Spaniards landed here in 1692, led by Sebastian Vizcaíno, they found Costanoan Indians living happily amid the abundant offerings of a benevolent sea.

Each spring and summer, upwellings of nutrient-rich cold water attract anchovies and squid, which in turn attract salmon, seabirds, and whales, making Monterey Bay one of the most diverse marine ecosystems in the world. In the first half of the 20th century, the world's largest sardine fishery supplied Monterey's rough-and-tumble Cannery Row, memorably portrayed in the eponymous novel by author John Steinbeck. By 1951 overfishing had sent the canneries into decline, and where boats bulging with sardines once thronged, tourists now gaze from waterfront restaurants.

BRILLIANT BRAINSTORM

Scientists first set up shop in Monterey Bay in 1892, when Stanford University erected a cluster of rough shacks on the waterfront that in time became the renowned Hopkins Marine Station. One evening in 1976, four young biologists from the marine station were sipping tequila when the conversation turned to the adjacent Hovden Cannery, which had been purchased by the university in 1972. As the four—Nancy Packard and her husband, Robin Burnett, along with Steven Webster and Chuck Baxter—pondered possible uses for the dilapidated cannery, one of them (no one can remember who) had a brainstorm. What if the cannery were turned into a new kind of marine center celebrating Monterey Bay's diverse ecosystem? The others in the group seized on the idea and together they proposed the creation of a public aquarium to David Packard,

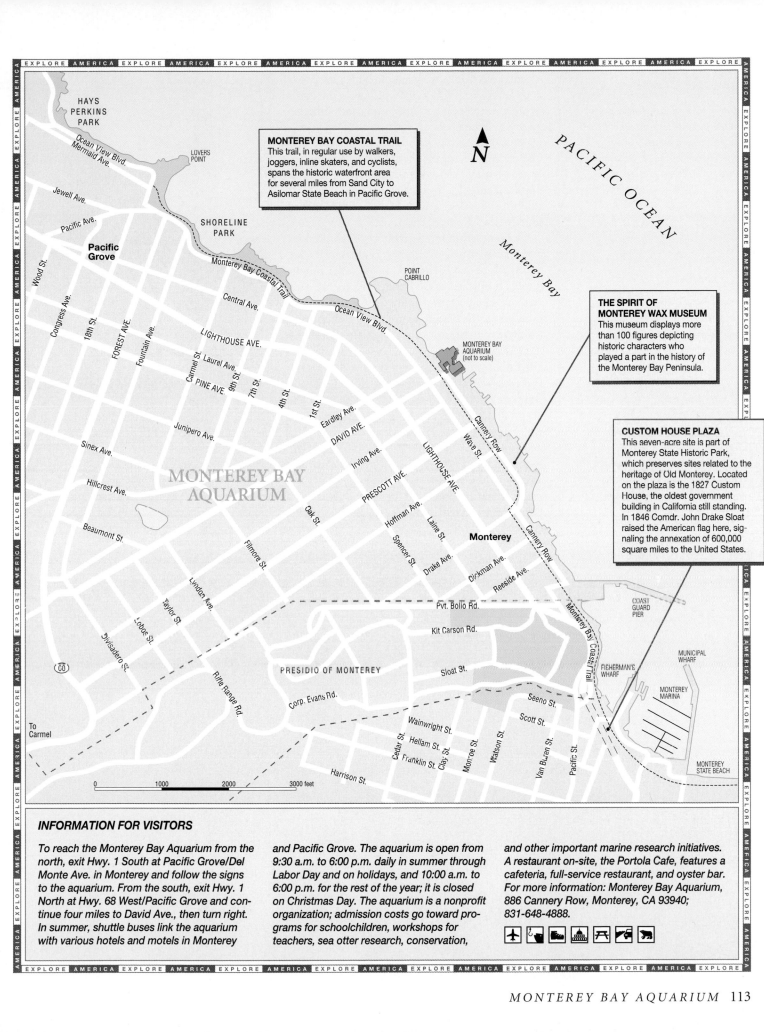

HAYS PERKINS PARK

Ocean View Blvd.
Mermaid Ave.

LOVERS POINT

MONTEREY BAY COASTAL TRAIL
This trail, in regular use by walkers, joggers, inline skaters, and cyclists, spans the historic waterfront area for several miles from Sand City to Asilomar State Beach in Pacific Grove.

Jewell Ave.

Pacific Ave.

SHORELINE PARK

Pacific Grove

Monterey Bay Coastal Trail

Wood St.

Congress Ave.

18th St.

Central Ave.

Ocean View Blvd.

POINT CABRILLO

PACIFIC OCEAN

Monterey Bay

THE SPIRIT OF MONTEREY WAX MUSEUM
This museum displays more than 100 figures depicting historic characters who played a part in the history of the Monterey Bay Peninsula.

FOREST AVE.

Fountain Ave.

Laurel Ave.

LIGHTHOUSE AVE.

Carmel St.

PINE AVE.

9th St.

7th St.

4th St.

1st St.

Eardley Ave.

DAVID AVE.

MONTEREY BAY AQUARIUM (not to scale)

Cannery Row

Wave St.

CUSTOM HOUSE PLAZA
This seven-acre site is part of Monterey State Historic Park, which preserves sites related to the heritage of Old Monterey. Located on the plaza is the 1827 Custom House, the oldest government building in California still standing. In 1846 Comdr. John Drake Sloat raised the American flag here, signaling the annexation of 600,000 square miles to the United States.

Junipero Ave.

Sinex Ave.

Irving Ave.

PRESCOTT AVE.

LIGHTHOUSE AVE.

Hillcrest Ave.

MONTEREY BAY AQUARIUM

Oak St.

Hoffman Ave.

Laine St.

Monterey

Beaumont St.

Filmore St.

Spencer St.

Drake Ave.

Dickman Ave.

Reeside Ave.

Cannery Row

COAST GUARD PIER

Lyndon Ave.

Taylor St.

Pvt. Bollo Rd.

Monterey Bay Coastal Trail

Lobos St.

Kit Carson Rd.

MUNICIPAL WHARF

Divisadero St.

68

PRESIDIO OF MONTEREY

Sloat St.

FISHERMAN'S WHARF

MONTEREY MARINA

Rifle Range Rd.

Seeno St.

Scott St.

To Carmel

Corp. Evans Rd.

Wainwright St.

Cedar St.

Hellam St.

Monroe St.

Watson St.

Van Buren St.

Pacific St.

MONTEREY STATE BEACH

Clay St.

Franklin St.

Harrison St.

0   1000   2000   3000 feet

INFORMATION FOR VISITORS

To reach the Monterey Bay Aquarium from the north, exit Hwy. 1 South at Pacific Grove/Del Monte Ave. in Monterey and follow the signs to the aquarium. From the south, exit Hwy. 1 North at Hwy. 68 West/Pacific Grove and continue four miles to David Ave., then turn right. In summer, shuttle buses link the aquarium with various hotels and motels in Monterey and Pacific Grove. The aquarium is open from 9:30 a.m. to 6:00 p.m. daily in summer through Labor Day and on holidays, and 10:00 a.m. to 6:00 p.m. for the rest of the year; it is closed on Christmas Day. The aquarium is a nonprofit organization; admission costs go toward programs for schoolchildren, workshops for teachers, sea otter research, conservation, and other important marine research initiatives. A restaurant on-site, the Portola Cafe, features a cafeteria, full-service restaurant, and oyster bar. For more information: Monterey Bay Aquarium, 886 Cannery Row, Monterey, CA 93940; 831-648-4888.

a leading electronics magnate and philanthropist—and Nancy's father. Packard was looking for a project that would bear his family's imprint. Although he was hesitant about an aquarium at first, knowing absolutely nothing about them, Packard and his wife, Lucille, began to tour existing aquariums. It soon became clear that there was a place for a marine center that combined state-of-the-art exhibitry with deep-sea research. In 1978 Packard founded the Monterey Bay Aquarium Foundation.

The foundation hired architects Linda Rhodes and Charles Davis of Esherick, Homsey, Dodge and Davis of San Francisco to design the aquarium. As the project grew, Rhodes became the project manager for the foundation, while Davis focused on the design. Committed to honoring Monterey's heritage, Davis integrated part of the old Hovden Cannery into the new structure, creating bold geometric compartments that mimic the cluttered intimacy of Cannery Row. Today the sunlit galleries, open ductwork, and restored boilers recall the site's earlier life as a sardine cannery.

The aquarium's engineers worked in synchrony with the ocean, first building a seawall near the high-tide line to create a work site for the main structure. At low tide, workers constructed forms heavy enough to withstand the pounding surf and poured the concrete that cured slowly underwater

during high tide. They then drilled through unstable coastal soils to anchor the main concrete piers to granite bedrock. Once this foundation was completed, the engineers went to work on the Kelp Forest and other tank exhibits.

Kelp are giant plants that attach their rootlike holdfasts to rocks on the seafloor. They can grow

up to 8 inches per day, and reach a height of about 100-plus feet at maturity. Their long tendrils provide a home for such crustaceans as snails and crabs, which in turn are eaten by fish, birds, and mammals. In order to exhibit the California coastline's giant kelp, engineers had to build what was then the tallest aquarium pool in the country.

SPECIAL DEMANDS

The aquarium pool's height created unique pressures on the tank floor and the need for a breakthrough design for the windows. Drawing on the wisdom of the ancient Greeks and Romans, the engineers mixed ashes, dust, and soot with the cement, creating a concrete strong enough to withstand the tank's load of 1,700 pounds per square foot. Expert concrete craftsmanship and long curing times reduced cracking to a minimum. After being poured into 18-inch-thick plates, the concrete was reinforced with an extremely dense six-inch mesh of steel rebar. To prevent corrosion from exposure to seawater, all the rebar was coated in epoxy—an innovation at that time in the aquarium field. Used primarily in highway decks and pavements, the epoxy-coated rebar demanded sensitive handling and precise placement. Once the concrete was poured, engineers applied a new calcium-based slurry sealant that formed crystals within the concrete itself, clogging the voids left by water that had evaporated during the drying process.

The tank's windows are revolutionary in design. In the words of Linda Rhodes, "The visitors had to feel like they were standing in a cathedral." But this cathedral-of-the-sea eschewed stained glass in favor of acrylic panes seven and a quarter inches thick. Softer than glass, acrylic is optically superior at larger sizes and stresses, and may also be polished to remove scratches. Manufactured in Japan, the panels were installed in a semicircle, creating a dramatic undersea theater.

Now all the exhibit needed was water—but not just any water would do. Aquarium scientists knew that the water itself would have to teem with life. Thankfully, they were able to pump seawater directly from the unpolluted Monterey Bay, an option denied many other marine centers. Two 16-inch intake pipes were laid 1,000 feet out into the Pacific, 55 feet below the water's surface. Scuba divers secured the pipes to the bottom with drilled rock anchors. At night unfiltered seawater pours through the exhibits, sustaining filter-feeding animals and sweeping in larvae and other life forms. During the day a complex filtration system clears the water for optimal visibility while still keeping the liquid "alive." A deftly calibrated surge machine and hidden water jets in rock walls maintain the flow of

new water to ensure that kelp and algae receive adequate nutrients. The water system worked perfectly and the kelp flourished, making the exhibit both a biological and architectural success.

Other live exhibits followed in swift succession. The Monterey Bay Habitats exhibit spotlights different local marine residents in another tank so huge that it allows sharks an unimpeded glide path of 90 feet. Outside the aquarium building, visitors can peer into the Great Tide Pool, where sea stars, urchins, and other creatures dwell in the intertidal zone in which water ebbs and flows twice daily over a low artificial rock wall. Inside, an unusual petting pool allows visitors to stroke bat rays whose stingers have been clipped. Nearby, sea otters—emblems of the central coast's threatened sea life—float on their backs and crunch crustaceans.

AMBITIOUS DREAM

The Monterey Bay Aquarium opened in 1984 to great public acclaim and a shower of awards for both its design and exhibits. But the work had only just begun. The next goal was to figure out a way to take visitors out into the boundless open ocean, a place few people, even divers, ever see. Certainly no aquarium had ever tried to display the open ocean, any more than a planetarium had tried to display deep space. Director of Live Exhibit Development Dave Powell defined the problem: "To interpret an environment without walls, you have to make the structure of the exhibit invisible." Powell engaged in "a fair amount of doodling" before one of his sketches, made on a cocktail napkin, seemed to contain the solution. For the Outer Bay exhibit, he envisioned a three-dimensional curve, "like the inside of an eggshell"—only this egg would be 90 feet

THE DRIFTERS

A life-size model of a killer whale greets visitors to the aquarium, right. In the aquarium's Drifters Gallery, above, sea nettles, purple-striped jellyfish, and comb jellyfish drift gently through the water just out of reach of awed spectators. Specially designed tanks hold the largest permanent collection of jellyfish species in the country.

long and 35 feet deep, and would contain a million gallons of water. Visitors would peer into the deep through the world's largest window, a single pane 15 feet tall and more than 54 feet long.

A huge amount of concrete could support the "eggshell" and its 8.5-million-pound embryo, but how would it affect the aquarium's foundation? Concrete could shape the shell itself, but could

The window posed another hurdle. It would have to be acrylic, enormous, and strong enough to resist incredible pressure. Only three companies in the world manufactured huge slabs of acrylic, and only the Nippura Co. of Japan had a viable plan: to construct five window sections in Japan and assemble them in Monterey. Nippura created an immense on-site "clean room" for chemical welding that

The sea is as near as we come to another world

craftsmen pull off the perfect pour? In the end, the designers realized that if the skin of the shell were a lightweight membrane, it could rest within a water-filled concrete tank—water would support water. This elegant solution led the designers to tap the expertise of William Kreysler, a specialist in composite materials, for his ideas on the best material for the shell. While Kreysler came up with a fiber-reinforced polyester laminate, computer modelers used aerospace design criteria to calculate the laminate's limitations under fluctuating conditions. In all, 13,000 pounds of resin and 7,000 pounds of fiberglass were used to reinforce the shell, one side of which was pressed flush against the viewing window. Lining the inside of the shell was a brilliant mosaic of 1.28 million blue glass tiles that produced the effect of endless ocean space.

involved superheating the sections inside an airtight cocoon for nearly a week. The result was a visually seamless sheet weighing 78,000 pounds. The window was soon put to the test: lifted by crane, it flexed once in midair—without cracking—and was fitted securely in place.

Several times each week the aquarium broadcasts "Live From Monterey Canyon," a video uplink from an unmanned submersible camera 6,000 feet below the surface. The ghoulish realm of gelatinous jellyfish, worms, and swirling detritus is the next frontier for the aquarium, which intends to be the first to display fragile deep-sea life forms. With more than 300,000 animals and plants already at home in 100 galleries and exhibits, the aquarium plans to greet the next century with yet more revelations from the unknown reaches of the sea.

ASTOUNDING SIGHT

In the introductory gallery of the Outer Bay exhibit, above, visitors are spellbound by the sight of some 3,000 anchovies, endlessly circling within the 15,000-gallon, doughnut-shaped exhibit tank.

A statue of Father Junípero Serra, above, stands near the gate of Mission San Carlos Borromeo del Río Carmelo. Father Serra founded the mission in 1770.

NEARBY SITES & ATTRACTIONS

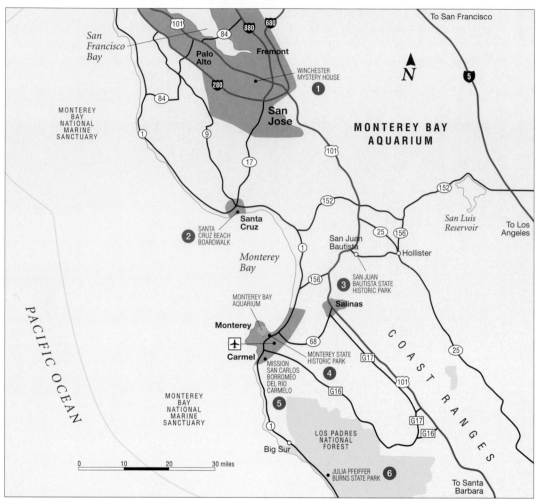

1 WINCHESTER MYSTERY HOUSE

After the deaths of her infant daughter and husband, Winchester firearms company heir Sarah Pardee Winchester consulted a psychic who convinced her that her personal tragedies were caused by spirits avenging those killed by the company's guns. The medium told her that she would be protected from danger if she built a lavish home for the souls of the kindly victims and kept enlarging it for as long as she lived. Using part of her $20-million inheritance, Sarah hired builders who eventually transformed the farmhouse into a sprawling mansion. She never consulted an architect; instead, she retired nightly to the Blue, or Seance, Room to consult the spirits for design instructions. In order to confuse evil spirits, some of the home's 2,000 doors lead to nowhere, and many of the 47 chimneys do not connect to the central chimney. Sarah maintained a staff of 18 servants to keep the house in impeccable condition, and to see to the wants of her unseen guests. Visitors who tour the 110 rooms open to the public will note the prevalence of combinations of 13, a number Sarah believed would bring her good luck: windows have 13 panes, sink drain holes have 13 holes, and the greenhouse has 13 cupolas. Located at 525 South Winchester Blvd. in San Jose.

2 SANTA CRUZ BEACH BOARDWALK

This historic boardwalk, the brainchild of promoter Fred W. Swanton, was patterned after those of Atlantic City and Coney Island. The boardwalk's 28 rides include the popular 1924 wooden roller coaster called the Giant Dipper and a 1911 Looff carousel, complete with 73 hand-carved wooden horses. Both of these rides have been designated as National Historic Landmarks by the National Park Service. Sunning, swimming, and sailing along the one-mile beach are also popular pastimes with locals and visitors alike. Located in Santa Cruz adjacent to the Monterey Bay National Marine Sanctuary.

3 SAN JUAN BAUTISTA STATE HISTORIC PARK

When the California missions were secularized by Mexican decree in 1834, settlers rushed to the area to obtain land grants. They built their homes beside the former missions. All the historic structures on the plaza of the old San Juan Bautista mission are part of this historic park. The adobe Monterey-style Castro House was erected in 1840 by the son of the town's founder, José María Castro. The house has been restored to reflect the lifestyle of the late 19th

century. The Plaza Hotel first served as a barracks in 1814, but in 1858 New Orleans restaurateur Angelo Zanetta added the second story, thus creating a thriving hotel that was a stopover point for travelers between Los Angeles and San Francisco. Located on Hwy. 156 in San Juan Bautista.

4 MONTEREY STATE HISTORIC PARK

This seven-acre site preserves several mid-19th-century adobe structures, most of which date from the time when Monterey ruled as the capital of Mexican California. The home of Thomas Larkin, who served as the U.S. consul to Mexico from 1843 to 1846, combines elements of Spanish Colonial and New England architectural styles. At the Robert Louis Stevenson House, visitors can see where the author lived while he courted a woman named Fannie Osborn and worked as a newspaper reporter for $2 a week. Inside the Custom House, the oldest standing government building in California, visitors can inspect trade goods from the 1840's. It was here, in 1846, that Comdr. John Drake Sloat raised the U.S. flag over California for the first time. Located at 20 Custom House Plaza in Monterey.

5 MISSION SAN CARLOS BORROMEO DEL RIO CARMELO

This mission, the second one built in the string of 21 missions that dot the coast of California, was founded in 1770 by Father Junípero Serra. The original church and dwellings were made of wood, but were later replaced with the more durable adobe. The church was rebuilt of sandstone in 1793. By 1834, when the missions were secularized, the area had been all but abandoned and gradually fell to ruin.

Today visitors can tour the restored church, with its Moorish-style tower, as well as Father Serra's spartan bedroom, complete with a bed made of boards. The priest is buried beneath the floor in front of the altar. Located at 3080 Rio Rd. in Carmel.

6 JULIA PFEIFFER BURNS STATE PARK

Perched on the wild southern coast of Big Sur, this 3,580-acre park features rugged sea bluffs and a waterfall. The park also preserves groves of chaparal and redwood trees. A variety of birds populates the park, including owls, hawks, vultures, cormorants, and pelicans. There are three hiking trails and two campsites. Located 35 miles south of Carmel on Hwy. 1.

The ocean licks at craggy rocks, above, in Julia Pfeiffer Burns State Park. The park was named for the daughter of John Pfeiffer, a late 19th-century rancher.

The restored colonnade, below, on the Santa Cruz Beach Boardwalk is a reminder of times past. The colorful stores that flank the colonnade sell everything from beach balls and cotton candy to wet suits and homemade chocolate.

GAZETTEER: *Traveler's Guide to Man-made Wonders*

Construction of Thomas Jefferson's head at Mount Rushmore, South Dakota.

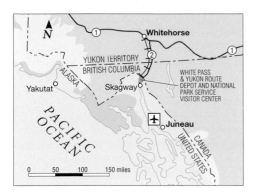

The White Pass & Yukon Route (WP&YR) that once linked Skagway, Alaska, to Whitehorse, Yukon Territory, was the northernmost railroad in the Western Hemisphere when it was built in 1898. The original 110-mile route climbed from sea level in Skagway to an elevation of 2,885 feet. Workers labored against all odds to build the railroad in just over two years. Operating along 40 miles of the original route today, the narrow gauge railway has grades as steep as 3.9 percent.

The WP&YR was born in 1898 during the Klondike gold rush, when thousands of fortune seekers made the grueling 40-mile trek from Skagway to Lake Bennett through the St. Elias Mountains. Here, the prospectors built their boats and headed out on the 500-mile journey down the Yukon River.

A study undertaken in 1896 indicated that the White Pass through the St. Elias Mountains was a viable railway route. Taking heed, a group of investors decided to explore the potential for a profitable railway to the Yukon, the destination of increasing traffic. In 1898 the consortium dispatched a survey party headed by the British aristocrat Sir Thomas Tancred.

In 1994 the WP&YR, above, was designated 1 of 22 International Historic Civil Engineering Landmarks.

After conducting a detailed survey, Tancred's experts decided White Pass was impassable. The project would have been scrapped but for a chance meeting in a Skagway hotel bar between Tancred and a young Canadian railway contractor named Michael Heney. Heney dismissed Tancred's doubts about the White Pass route. "Give me enough dynamite . . ." he declared, "and I'll build a road to Hell!" The pair talked long into the night, and by morning Heney had landed the railway deal of a lifetime.

CONSTRUCTION BEGINS

Construction from Skagway began on May 28, 1898, fueled only by men, horses, and 450 tons of explosives. Cash-strapped and Klondike-bound, an itinerant workforce of 30,000 endured chilling temperatures, heavy snows, and savage terrain to hack out the railway.

The line advanced past glaciers and waterfalls and slowly crawled through the St. Elias Mountains, inching along trestle bridges and clinging to the rims of yawning precipices. Suspended by ropes from sheer cliffs, work crews wielded picks to chip blast holes into the rock, then planted black powder to tear through the virgin granite. In February 1899, nine months after its start, the railroad line reached the summit at White Pass. By July 1899 the track curved along the shores of Lake Bennett.

Workers encountered numerous obstacles along the route, including a 198-foot-long stretch of frozen quicksand. At Lewis Lake, workers unintentionally lowered the water level by 70 feet—59 feet more than required to accommodate the grade—and were forced to construct two additional bridges. Moreover, costs climbed as fast as the grade, rising from an initial estimate of $1,570,000 to a final cost of $10 million.

As the southern work gang blasted its way through the difficult White Pass, a northern crew worked doggedly toward Whitehorse. On July 29, 1900, the rails met at Carcross, Yukon Territory.

Ironically, by the time the ceremonial spike was driven at Carcross to commemorate the completion of the route, the stampede to the Yukon had waned and the gold rush was all but finished. Nevertheless, the route proved invaluable for hauling passengers and freight to the Yukon. It also became a chief supplier for the army's Alaska Highway construction project during World War II; later still, it gained a global reputation as an International Historic Civil Engineering Landmark.

Today a rolling museum of vintage parlor cars, powered by a restored locomotive, carries passengers the 40 miles from

The construction of the WP&YR, above, has frequently been called the toughest railway job ever undertaken.

Skagway to Lake Bennett, pausing at Canadian customs at Fraser, 8 miles past the Canadian border at the summit of White Pass. Engineering relics seen along the winding route include a steel cantilever bridge that was the highest of its type in the world when it was completed in 1901. Passengers can spot the weblike trestle rising 215 feet above Dead Horse Gulch, which is still littered with the bleached bones of some of the 3,000 pack animals driven to their death during the feverish rush to the goldfields.

FOR MORE INFORMATION:

White Pass & Yukon Route, P. O. Box 435, Skagway, AK 99840; 907-983-2217 or 800-343-7373.

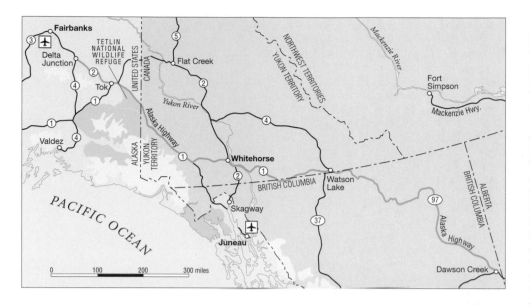

Signpost Forest, right, near Watson Lake, Yukon Territory, was initiated by a construction worker who tacked up a sign showing the distance to his hometown. The collection has been growing for more than 50 years, and includes 12,000 signs from such far-flung towns as Brno, Czech Republic, to Dumplin, Tennessee.

The mystique of the Alaska Highway, the sole overland supply route in Alaska, has endured since its construction was deemed a military necessity more than 50 years ago. Slicing through endless stands of evergreens, the two-lane roadway runs a full 1,420 miles between Dawson Creek, British Columbia, and Delta Junction, Alaska.

The backbreaking work began in March 1942, shortly after the Japanese bombing of Pearl Harbor made a West Coast invasion seem imminent. The United States and Canada, fearing that their Pacific sea lanes would be cut off by the Japanese, joined forces to ensure delivery of vital supplies and equipment via an overland route. In return for rights-of-way through Canadian territory, the Americans agreed to pay for the construction of the road and turn the Canadian portion of the highway over to the Canadian government after the war.

The U.S. War Department determined the route, which followed a chain of existing airfields, winter roads, old Indian trails, and rivers. Trappers and natives guided surveyors through tundra and thick forests, muskeg swamps, plateaus, and steep hills etched with glacial streams. Once the route was laid out, more than 11,000 troops and 16,000 civilians arrived in the north and started to work on the project. Dubbed the Alaska–Canada Military Highway, or Alcan,

the roadway took eight arduous months to complete. Laborers constructed an average of eight miles a day through the wilderness. They also built more than 8,000 culverts—underground channels beneath the roadway—and 133 log and pontoon bridges, which were later replaced with sturdy bridges made of steel.

Workers drove bulldozers and other heavy equipment as far as tracks and trails would allow. They then took the massive machines apart so that bush pilots could fly the pieces farther north. Equipment was constantly in short supply and breakdowns were common.

Workers labored seven days a week, enduring swarms of mosquitoes and black flies in summer, and frigid temperatures in winter. They slept in canvas tents heated only by potbellied woodstoves that barely protected them from frostbite. Not surprisingly, these conditions contributed to a rapid turnover in the labor force that, according to Lee Carman, a former construction supervisor, made the work all the more difficult. "The only thing that used to get my goat," recalled Carman, "was some of those doggoned guys just wouldn't stay. There was times when you could say we had one crew working, a crew going, and a crew coming."

Despite such problems, the work continued at a breakneck pace, spurred in June 1942 by reports of the Japanese invasion of two islands in the Aleutians. Crews working from the east and west converged at Contact Creek on September 25, 1942.

The highway officially opened on November 20, 1942. Loggers, gold miners, truckers, and passenger traffic now had a route to the north. Many people who came to visit ended up staying, and small towns began to line the highway.

This landscape of windswept tundra and rugged mountain ranges is patched with designated wilderness areas, including the Tetlin National Wildlife Refuge, which preserves 924,000 acres of unique waterfowl habitat. Other residents of the refuge's river basins and marshlands include moose, caribou, grizzlies, and black bears, while arctic grayling, whitefish, and trout are abundant in the region's lakes and rivers.

Road crews continue to wage an ongoing battle against the elements. The all-weather route is marked by potholes, deteriorated shoulders, stretches of gravel, and frost heaves—a rippling effect in the pavement caused by cycles of freezing and thawing. But the poor road conditions and lonely stretches of highway have little discouraged summer adventurers, who relish traveling along the historic highway that broke Alaska's shell of isolation.

FOR MORE INFORMATION:
Alaska Division of Tourism, P.O. Box 110801, Juneau, AK 99811-0801; 907-465-2012.

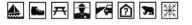

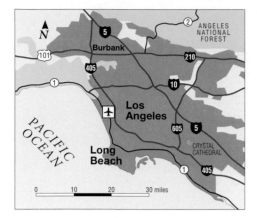

Twelve thousand panes of tempered, silver-colored glass make this house of worship the largest all-glass structure in the world. Established by the Iowa-born pastor Robert Harold Schuller, this Protestant church is the site of Schuller's television ministry, which reaches more than 20 million viewers worldwide.

Robert Schuller, a Reformed Church in America minister, first came to Garden Grove in Orange County with his wife, Arvella, in 1955. In the early days, Sunday services were conducted at a drive-in theater, with Schuller preaching from the roof of the snack bar and Arvella accompanying him on an organ.

Arvella spearheaded a campaign to raise the funds necessary to purchase a special organ for her husband's church, which was under construction. The Hazel Wright Organ, named after the late Chicago benefactor who donated the instrument and an endowment for its upkeep, was installed in the Crystal Cathedral in 1981—one year after the cathedral was dedicated. The organ, designed by the virtuoso organist Virgil Fox, is made of elements from a 1962 Aeolian-Skinner instrument from New York's Avery Fisher Hall and a 1977 Ruffatti organ that was used in the former sanctuary of the Garden Grove congregation. The Hazel Wright Organ has 16,000 pipes in its 287 ranks of pipes, and is played from a five-manual console—the largest drawknob console ever built—that is mounted on a movable platform.

The ambitious design of the Crystal Cathedral was the brainchild of architect Philip Johnson, famed for such edifices as the Seagram Building and the AT&T headquarters, both in New York City. But it was not Johnson's first all-glass building: in 1949 the renowned architect had constructed the single-room Glass House on his estate at New Canaan, Connecticut.

Some of the dimensions of the building match those of many of the great stone

The spacious chancel area of the Crystal Cathedral, above, contains one of the five largest organs in the world.

cathedrals of Europe—a length of 415 feet and walls soaring to 128 feet. Its 207-foot width, however, is much greater than that of its Old World cousins. The glass panes of the walls and roof are held in place by a framework of white steel trusses. The glass-and-steel construction gives an impression of lightness and makes the cathedral seem even larger than it actually is.

The building's layout resembles a four-pointed star with the chancel, altar, and pipe organ placed in the northern point. The sanctuary seats 2,736 people, with room for another 1,761 on the main floor. There are still more seats in the east, west, and south balconies, which form the other points of the star-shaped design. The massive concrete columns that support the balconies are hinged in two places—at the balcony and foundation levels—to allow for movement in the event of an earthquake.

On September 16, 1990—the cathedral's 10th anniversary—a 236-foot-high bell tower was dedicated before a crowd of thousands. The bell tower, also designed by Philip Johnson, is constructed of highly polished steel prisms. The 52-bell carillon was named in honor of Arvella Schuller.

SPANISH MARBLE AND GIANT TVS
The 185-foot-long main chancel, which can hold more than 1,000 singers and instrumentalists, is built of Spanish marble; it features a granite altar table and pulpit, as well as a 17-foot-tall cross. A giant indoor television screen provides close-up views of the choir. Behind the pulpit, two

90-foot-tall doors open to admit the morning breezes. Outside, worshipers can view the services on another huge screen—a high-tech reminder of the church's early drive-in days in Garden Grove.

FOR MORE INFORMATION:
Crystal Cathedral, 12141 Lewis St., Garden Grove, CA 92840; 714-971-4000.

The 12-story Crystal Cathedral, above, is set within a 21-acre campus that also includes an arboretum, fountains, and the peaceful Walk of Faith Garden.

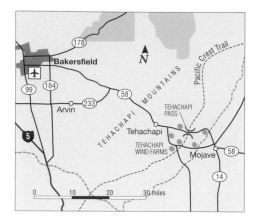

A quiet corner of Kern County, not far from the small southern California towns of Tehachapi and Mojave, may seem an unlikely setting for a revolution. But the rolling hills of Tehachapi Pass are the site of an exciting vision of a future free of dependence on fossil fuels. Here, orderly rows of 5,000 wind turbines silently turn in the wind. Operated by 12 private energy companies, the wind turbines annually supply the California electricity grid with enough power for 500,000 residential customers, or about 1 percent of the state's electricity needs.

In bygone days, wind power was harnessed to propel sailing ships and turn the sails of mills used to grind grain or pump water. However, the large-scale use of the thousands of wind turbines to provide an infinite renewable, non-polluting resource dates from the 1973 energy crisis, when the price of imported oil increased astronomically.

WINDY PASS
Tehachapi Pass is one of the windiest spots in the world. During the day the sun heats up the dry air of the Mojave Desert to the southeast. As the warm air rises, more air is drawn through the pass from Central Valley to the northwest to replace it. The greater the pressure difference between the two areas, the faster the wind. The average wind speed in the Tehachapi Pass is 14 to 20 miles an hour, but wind speed varies greatly depending on the terrain, season, and time of day. Most turbines in the pass spin at wind speeds of between 15 and 50 miles an hour. In general, the wind is greatest in the late afternoon, when the fields of spinning turbines are an awe-inspiring sight.

The turbines are of several different types. The majority resemble propeller-driven aircraft engines and are mounted on high steel masts, or towers, ranging from 80 to 200 feet in height. There are also several hundred turbines of the Darrieus type mounted

Wind turbines at Tehachapi Wind Farms on Tehachapi's hilltops, above, produce one of the cheapest sources of electricity in the world, while at the same time existing in harmony with their surroundings.

on Cameron Ridge, close to the spot where the Pacific Crest Trail crosses Highway 58. Standing about 100 feet high and 50 feet wide, these distinctive turbines resemble eggbeaters, and can accept wind from any direction. Sensors tell the turbines when to start up and, if the winds are too strong, when to switch off. Wind vanes permit most machines to change direction as the wind shifts. Management of the more modern machines can be controlled by an operator at a central computer. Although the turbines take up a great deal of land, crops can grow and livestock can graze peacefully around the turbine masts.

Vantage points in the Tehachapi–Mojave Wind Resource Area allow visitors to see the turbines in action. At Sea West Tehachapi on Oak Creek Road, just west

of the town of Mojave, visitors can enjoy the sight of Danish and Japanese wind turbines. The Pacific Crest Trail offers dramatic views of the wind turbines and the outlying Antelope Valley and Mojave Desert, but take note: the wind farm personnel advise visitors not to leave the trail and approach the turbines. Each year the Tehachapi Wind Fair celebrates the area's blustery bounty with stalls and exhibits that enlighten visitors about the benefits of renewable and alternative energy sources.

FOR MORE INFORMATION:
Kern Wind Energy Association, P.O. Box 277, Tehachapi, CA 93581; 805-822-7956.

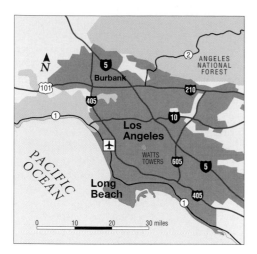

A collection of fantastical sculptures in the Watts District of Los Angeles stands as a monument to the ingenuity and perseverance of one man—Simon Rodia. For 33 years Rodia labored single-handedly toward the realization of his artistic vision: a series of delicate spires made of steel and mortar and encrusted with colorful bits of tile, shell, pottery, and glass. One tower is 99.5 feet high, the tallest slender reinforced concrete column in the world. Although officially called the Towers of Simon Rodia, the structures are referred to as the Watts Towers because they are located in the Watts neighborhood of the sprawling city.

Rodia, born in Italy in the late 1870's, immigrated to the United States with his family during the 1890's. Little has been documented about his early life, but it is known that Rodia's family settled in Pennsylvania, where an older brother supported the family by working in the coal mines. Eventually Simon moved to the West Coast, where he worked in rock quarries, railroad camps, and later became a construction worker and tile setter.

After Rodia purchased a lot at 1765 East 107th Street in Los Angeles in 1921, he began to construct a series of skeletal towers that he called *Nuestro Pueblo*, meaning "Our Town." He worked without the benefit of scaffolding, machine equipment, bolts, rivets, welds, or drawing board designs. His tools were those of a tile setter: pliers, a hammer, and wire clippers, which he carried in a window washer's belt.

Each of Rodia's sculptures is adorned with irregularly shaped materials, including shards of mirrored glass, early 20th-century American ceramics, and a portion of rare 19th-century hand-painted Canton ware. Rodia interspersed his mosaics with hand-drawn designs of flowers, suns, hearts, and spirals, which he imprinted in the mortar.

Over the years Rodia's creation grew into nine major sculptures, including three towers measuring 55 feet, 97.5 feet, and 99.5 feet tall, respectively, and two smaller towers, 25 feet and 14 feet tall. One piece, titled *The Ship of Marco Polo*, is topped by a 28-foot spire, while another tower, referred to as *The Gazebo*, includes a 38-foot spire, a circular bench, three birdbaths, and a surrounding garden. The ruins of Rodia's house are located near his soaring artworks. All that remains of the structure is a decorated front facade and a mirrored canopy entranceway with a 25-foot-high chimney. The south side of Rodia's lot is bordered by a 240-feet-long scallop-edged wall.

SOMETHING BIG

In 1954, when the sculptor was in his mid-seventies, he deeded his property to a neighbor and retired to be near his family in Martinez, California, where he died on July 16, 1965. Although the significance of Rodia's sculptures remains unclear, the artist reputedly once said, "I had in mind to do something big, and I did it."

In the late 1950's the Department of Building and Safety ordered that Rodia's pieces be demolished. Fearing the loss of this distinctive monument to American folk art, a group of citizens formed the Committee for Simon Rodia's Towers in Watts; by collecting money and signatures, they fought to save Rodia's work. The committee also devised an engineering test that proved the towers were structurally sound.

The towers are now a unit of the California State Park System and are currently administered by the Cultural Affairs Department of Los Angeles. Designated a National Historic Monument in 1990, Rodia's slender spires are undergoing restoration to repair large cracks in the concrete that were caused by rain. The towers are also being reinforced to withstand potential earthquakes.

The upper portions of the sculptures are visible from the Watts Towers Arts Center, which is located next to the towers. The center sponsors tours of the site, lectures, and changing art exhibits, and offers studio workshops for both teachers and schoolchildren. Each September the center hosts the Simon Rodia Watts Towers Jazz and Arts Festival, which showcases the work of musicians who perform jazz, gospel, and rhythm and blues. The spirit of Rodia's creative endeavor extends to the annual Watts Towers Day of the Drum Festival, which features drummers of all stripes: Afro-Cuban folk musicians, Afro-Brazilian percussion experts, American jazz artists, and Japanese taiko drummers.

A close-up of one of Rodia's towers, above, displays the inlaid tiles and the designs that the self-taught sculptor drew in the mortar while it was still wet. The spires are made of wire mesh smeared with builder's cement.

FOR MORE INFORMATION:
Cultural Affairs Dept., City of Los Angeles, 433 South Spring St., 10th Floor, Los Angeles, CA 90013; 213-485-2433.

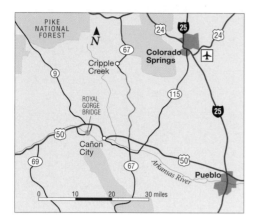

The Arkansas River tumbles down from the mountains of central Colorado on its way eastward to the Mississippi. Over a period of many millions of years, the river carved out a path through the hard granite of Fremont Peak, now known as the Royal Gorge. The Royal Gorge Bridge—the world's highest suspension bridge—spans the 1,000-foot-deep chasm. A dramatic sight, the bridge is suspended 1,053 feet above the roiling blue-green waters of the Arkansas River.

Constructed in 1929, the Royal Gorge Bridge has become one of the West's most popular attractions. From the wooden deck, visitors can savor wide-angle views of the gorge and surrounding mountains. An incline railway—the steepest in the world—descends the side of the gorge to river level, while an aerial tramway provides another perspective of the area.

THE ROYAL GORGE WAR

Looking down into the V-shaped gorge, it seems incredible that engineers could ever have built a railroad through the chasm. At the gorge's narrowest point, railroad engineers had to run the tracks over a hanging bridge suspended from steel beams driven into the rock walls on either side.

Odd as it may seem, it was not the engineering of the route that raised the most problems, but rather the struggle between the two railroad companies that wanted it built. In the 1870's the discovery of silver near Leadville led the Santa Fe and Denver & Rio Grande railroads to compete over control of the route to the mining camps. The short-lived Royal Gorge War, as the bitter rivalry was known locally, erupted into repeated violence as the companies vied for the right to lay tracks through the gorge. During the day work crews would lay sections of track along a route determined by their particular railroad company. Then, after dark, the workers would sneak over to the tracks built by their opponents and

blow the tracks up. The conflict finally ended up in court. The victor was the Denver & Rio Grande, which completed the tracks to Leadville by July 1880. In compensation for its earlier work, the Santa Fe company received $1.4 million.

The spectacular rail journey through the gorge drew large numbers of sightseers,

Cañon City derives a significant portion of its municipal revenue from tourists who visit the Royal Gorge Bridge. In 1929 the bridge cost just over $350,000 to build. Today it would cost in the neighborhood of $10 million.

especially after the rise of the automobile made the site easier to reach. In 1907 the federal government deeded the Royal Gorge to nearby Cañon City for use as a municipal park. It was in the 1920's that Cañon City decided to draw even more tourists to the area by building a bridge over the gorge.

Construction on the ambitious project began on June 5, 1929. To support the 150-foot-high steel towers of the bridge, workers built massive stone and concrete abutments on the north and south rims of the gorge. When the abutments and towers were complete, two cables were slung across the gorge. Each cable consists of 2,100 individual wires that are anchored to steel pins driven into the solid granite walls. The cables are buried in trenches excavated by the workers, who had used the rubble to build the abutments. To ensure that the cables were securely anchored, the trenches were filled with concrete and reinforced with steel bars. The crews then attached vertical suspenders to the cables and bolted the suspenders to the steel girders for the roadway. A plank deck was attached to this steel framework and protective railings were added along the sides.

The Royal Gorge Bridge was dedicated on December 6, 1929. In the five months that it took to build the bridge, not a single worker lost his life and no major accidents took place. An incline railway was built immediately after the bridge was completed. The tracks run at a 45-degree angle over steel I-beams that are set into the walls of Telephone Gulch, a side gorge located just west of the bridge on the north rim. The ride to the bottom of the canyon takes five minutes; the 30-person cars are hauled up and down by two .75-inch hoist cables and a 1.75-inch safety cable; if one cable were to break, either of the other cables could support the entire weight of the car.

To prolong the life of this one-of-a-kind bridge, vehicular traffic has been reduced and buses and recreational vehicles are no longer permitted to make the crossing. The most spectacular addition to the gorge was the installation of a 2,200-foot aerial tramway in 1968. The 35-passenger tram cars whisk visitors on a five-minute, gravity-defying ride over the chasm. Although the view of the gorge from the tramway is spectacular, the man-made bridge is even more breathtaking.

FOR MORE INFORMATION:

Royal Gorge Bridge, P.O. Box 549, Cañon City, CO 81215; 719-275-7507.

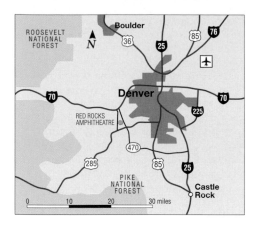

Doors to the backstage area, left, lead to dressing rooms that highlight the exposed natural sandstone and rock formations.

I n 1911 famed soprano Mary Garden rode into the mountains 17 miles from Denver and stopped to sing a few songs in a natural theater sculpted from rock, an area known as Red Rocks. After singing "Ave Maria," Garden marveled, "Never in any opera house the world over have I found more perfect acoustic properties." She declared that someday people would gather here to listen to masterpieces.

Garden's prophecy has been fulfilled. Ever since the Red Rocks Amphitheatre was completed in 1941, thousands of visitors have flocked here to listen to world-class performers ranging from rhythm and blues guitarist B. B. King and the Colorado Symphony Orchestra to Bruce Springsteen and U2, who filmed their rock video *Under a Blood Red Sky* at the site in 1983.

NEAR PERFECTION
Millions of years ago the entire region was engulfed by an inland sea. After the waters receded, sand from the seafloor hardened into stone, which was then thrust into near-vertical structures by the geological upheavals that formed the Rockies. The massive rock formations that cradle the stage are streaked with shades of red—from terra-cotta and vermilion to maroon. The colors are produced by varying quantities of iron oxides in the sandstone.

John Brisben Walker, founder of *Cosmopolitan* magazine, recognized the site's potential as a concert venue when he purchased it in 1909. Financial troubles forced him to sell the property to the city of Denver in 1927, but his dream was revived by George Cranmer, then manager of Denver Parks and Improvements. Having visited the ancient Greek outdoor theater at Taormina in Sicily, Cranmer envisioned

Red Rocks Amphitheatre, right, features near-perfect acoustics and open-air seating for up to 9,450 patrons.

a similar amphitheater here. He persuaded Denver mayor Ben Stapleton to support the project, then commissioned local architect Burnham Hoyt to draw up plans, stipulating that the design preserve the area's natural beauty. Labor and materials for the project were provided by Franklin D. Roosevelt's Civilian Conservation Corps (CCC), a Depression-era program designed to provide work for the unemployed.

The theater now boasts a concrete stage that measures 60 feet by 70 feet and a roof designed to withstand winds of 100 miles an hour. Spectators can see the lights of the Denver skyline peeking over the top of the stage, which is backed by a wall of rock. The rake of the seating ensures that each spectator has an unimpeded view of the stage.

Designated a National Historic Site in 1990, the amphitheater was also voted best outdoor concert venue for five years in a row by *Pollstar* magazine. Perhaps the venue's only drawback is its high elevation: during the Beatles' 1964 concert, members of the Fab Four needed frequent hits of oxygen throughout their performance. Nevertheless, Red Rocks' natural acoustics have put it on the map for music lovers from around the world.

FOR MORE INFORMATION:
Red Rocks Amphitheatre, City and County of Denver, Theatres and Arenas Division, 1380 Lawrence St., Suite 790, Denver, CO 80204; 303-640-2637.

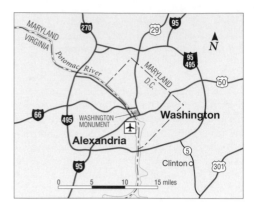

I t stands arrow-straight, honest, and true—just like young George Washington, who could not tell a lie when he cut down the cherry tree. The dignified Washington Monument commemorates the achievements and legacy of the man whose leadership in peace and war placed the government of the infant republic on firm footing. The obelisk, aligned precisely with the cardinal directions, towers just over 555 feet above the verdant National Mall in the nation's capital.

The project took more than a century to complete, during which time it was bedeviled by a chronic lack of funds and political bickering. As early as 1783 the Continental Congress decided that a statue of the great man mounted on horseback should be erected. Washington himself chose the site. Unfortunately, opposing opinions on the design of the obelisk that was eventually chosen delayed its construction until July 4, 1848, when the cornerstone was laid.

In the interim, a monument society, formed in 1833, set the ambitious—and unrealistic—goal of raising $1 million by public subscription. In 1855 control of the project was seized by the anti-immigration and anti-Catholic group known as the American Party. The group, which was nicknamed the Know-Nothings, gained control of the monument society at an illegal election held at an annual meeting of the society on February, 22, 1855; they even went as far as throwing the stone donated by the pope of that time into the Potomac River. Work came to a standstill for several years, and in 1861 the Civil War intervened. It was only in 1876—the centennial year—that Congress voted the funds required to finish the project and appointed military engineer Lt. Col. Thomas Casey to supervise construction.

SIMPLICITY OF FORM

The plain obelisk that stands in the mall is quite different from the original plan for the monument. A design competition was held in 1836. The winning entry, submitted by Robert Mills, called for a 600-foot shaft decorated with inscriptions and scrolls and rising from the center of a circular pillared structure that would house statues of Revolutionary War heroes. After Casey took over construction, the design was altered to a hollow obelisk with no decoration—both for esthetic reasons and in the interest of completing the project.

Obelisks, tapering four-sided shafts with pyramidal tops, originated in ancient Egypt and were popular in Roman times. But ancient obelisks were little more than 100 feet high. To determine the correct proportions for the Washington Monument, Casey corresponded with the American government's representative in Rome, who advised him that the height of an ancient obelisk was 10 times the width of its base. Because the monument's base was 55 feet wide, Casey adjusted the height of the final design from 600 to 555 feet. The height of the monument is precisely 555 feet, 5 and one-eighth inches, making it the world's tallest freestanding masonry structure.

Casey had to strengthen the foundation before workers could lay the rest of the stone. Egyptian obelisks were carved from a single block of stone, but the Washington Monument is built of thousands of blocks of Maryland and Massachusetts marble, laid without mortar in an arrangement called a Flemish bond. The original builders had used Maryland marble for the first 150 feet of the monument. Casey tried to match the stone with marble from a Massachusetts quarry, but when the company defaulted on its contract, he went back to the Maryland quarry. Although the new marble was cut from the same vein, it came from a different level. Throughout the years the stone had been exposed to the elements, making the difference in the marbles apparent; this accounts for the lighter color of the shaft's bottom third.

When the walls reached 500 feet in height, the pyramidal top was hoisted into position and covered with a solid aluminum cap and five platinum lightning rods. The 9-inch-tall, 100-ounce cap was the largest piece of aluminum ever cast at the time, and it was exhibited in Tiffany's jewelry store in New York before being placed on the monument. People who came to see the cap often leaped over it, later bragging that they had stepped over the top of the Washington Monument.

The Stars and Stripes encircle the Washington Monument, right, one of the nation's most visited landmarks.

Once the cap was on, the official opening ceremony took place on February 21, 1885, when Pres. Chester Arthur dedicated the towering obelisk.

Today more than 1 million people visit the monument every year, making it one of the capital's most popular sites. Most people take the 70-second elevator ride to the observation area, where they enjoy bird's-eye views of the White House, the Capitol, the Lincoln Memorial, and other city landmarks. Ranger-led tours on the weekends guide visitors down 897 steps and past 192 memorial stones presented to the project by states, countries, cities, clubs and organizations, and private individuals. (The first memorial stone came from the state of Alabama in lieu of a cash donation.) On a more mundane note, rangers point out the 19th-century graffiti in the lobby.

FOR MORE INFORMATION:
Washington Monument, National Capital Parks Central, 900 Ohio Dr. SW, Washington, DC 20024; 202-426-6839.

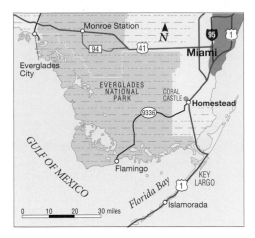

Ed Leedskalnin's coral throne room, above, contains a 5,000-pound throne for himself and several smaller ones, supposedly for his Sweet Sixteen, her mother, and a child. Visitors often use the Moon Pond, left, as a wishing well.

Coral Castle and its gargantuan sculptures are the remarkable achievement of a man who measured his heartbreak in stone. Made of more than 1,100 tons of rock, the castle walls, tower, and sculptural carvings were built single-handedly by Ed Leedskalnin, a Latvian immigrant who weighed a mere 100 pounds and stood just over 5 feet tall.

Leedskalnin spent 25 years building his stone tribute to a woman whom he called his Sweet Sixteen. Although Leedskalnin never revealed his engineering techniques to anyone, the diminutive builder somehow managed to move and sculpt slabs of coral bedrock that were on the scale of the stones used to build the Great Pyramid of Giza. He claimed, in fact, that he used the same methods that were employed to construct the pyramids to build what is considered to be one of the finest examples of massive stone construction in the nation.

Born in a small village in Latvia in 1886, Leedskalnin was 26 years old when he was jilted by his 16-year-old sweetheart, Agnes, on the eve of their wedding night. Devastated, he embarked on an odyssey through Western Europe and Canada. He was working as a logger in the state of Washington when he was diagnosed with tuberculosis.

Leedskalnin moved south, and in 1918 he spent $12 on an acre of land in Florida. Working by night to avoid detection, he excavated and carved tons of native coral bedrock, a substance so hard it can break diamond-tipped saws and drill bits. Yet the only tools Leedskalnin used to build his castle were implements made of discarded objects, including crude winches, iron wedges fashioned from truck springs, and a ladder built from pieces of an old car. Also remarkable was the fact that the builder, who had only a fourth-grade education, designed and built an electrical generator that served as his only source of power.

In 1936 Leedskalnin moved his colossal sculptures to a 10-acre plot of land he had bought near Homestead. He loaded and unloaded the trailer in secret. When he reached his new home, which he dubbed Rock Gate Park, Leedskalnin erected an eight-foot-high wall around it.

UNUSUAL ABODE
The sculptor's living quarters and the two-story main tower were constructed using four- to nine-ton coral rocks. The completed building weighs an amazing 243 tons. The house is modestly furnished with unusual objects, including a bed made of boards wrapped in burlap; a hanging chair made from a horse's harness, scrap metal, and bicycle parts; and a kerosene stove with a food box suspended above.

Leedskalnin's sculptures are scattered around the grounds. Among the impressive monuments is a 9-ton gate with a massive stone slab door so perfectly balanced that a child can open it with the touch of a finger; the door measures 80 inches wide and 21 inches thick. The castle's largest sculpture, the Great Obelisk, weighs more than 28 tons and stands 25 feet high—taller than the Great Upright monolith at Stonehenge, England.

One of the stone tables is hewn into the shape of the state of Florida, while another is carved into a 5,000-pound heart, which —according to *Ripley's Believe It or Not!*— is the world's largest valentine. The Moon Pond features two massive crescents depicting the first and last quarters of the moon. A 23-ton fountain represents the full moon, and a sundial tells standard time accurately.

In 1936 Leedskalnin wrote, "We always strive for perfection. We are only one-half of a perfect whole; man is the bigger and stronger half and the woman is the other." He remained unlucky in love until his death in 1951. Nevertheless, his legacy continues to delight visitors and baffle scientists and engineers, who have yet to uncover how he managed to build his astonishing monument to a lost romance.

FOR MORE INFORMATION:
Coral Castle, 28655 South Dixie Hwy., Homestead, FL 33030; 305-248-6345.

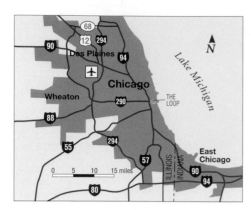

The Monadnock Building, right, completed in 1893, was one of the last of the great masonry-wall harbingers of the skyscraper.

After the Great Fire of 1871 destroyed more than 17,000 wooden buildings in Chicago, the city's residents cleared away the rubble and rebuilt. Instead of lamenting the past, Chicagoans looked boldly to the future and erected tall, awe-inspiring buildings that would proclaim the city's growing importance as the hub of the railroad network in the Midwest and the home of more than 500,000 people.

The buildings constructed during the late 1880's and 1890's are in a section of Chicago encircled by an elevated train system—a district known, not surprisingly, as the Loop. Several of these buildings represent important milestones in the development of skyscraper technology. One key innovation that enabled architects and engineers at the time to construct multistory structures was the invention of load-bearing steel skeletons that sit upon a concrete foundation. Another was safe elevators, which could whisk passengers and goods from floor to floor.

ARCHITECTURAL TOUR

Today a walking tour of the Loop showcases some of Chicago's splendid architectural heritage. The 16-story Monadnock Building on West Jackson Boulevard is the world's tallest masonry-walled building. It has six-foot-thick brick walls at the base that taper at the top and is distinguished by its almost total lack of exterior decoration. John Wellborn Root was responsible for the northern half, which was completed in 1891; the southern half, which had a steel skeleton frame, was designed by the firm Holabird & Roche and was completed in 1893. For years it was considered to be on the cutting edge of design: portal bracing, previously used only on bridges, was used to support the vertical columns and walls against the force of high winds.

Working with his partner, architect Daniel H. Burnham, Root was also instrumental in the creation of the 1888 Rookery Building on South LaSalle Street. The 11-story structure, a precursor to the skyscraper, got its name from the hundreds of pigeons that roosted on the temporary city hall that previously had stood on the site. The building's masonry load-bearing walls in its street facades, combined with cast-iron columns and steel beams, mark it as a transitional design for skyscrapers. Built around a central court surmounted by a skylight, the building was described as "a thing of light" by contemporaries. Attractions include Romanesque entry arches, an elegant cast-iron oriel staircase, and a sparkling gold and ivory atrium designed by architect Frank Lloyd Wright when the court was renovated in 1907.

Burnham's firm was also involved in building the 18-story Fisher Building, which represented the next step in skyscrapers. Constructed with a steel frame and covered with lightweight fired clay called terra-cotta, the 1896 building on South Dearborn Street was one of the lightest ever made. The steel frame proved economical: although the actual material was expensive, construction was quick and cheap. At the time, it set a record for the quickest assembly of a steel-frame building: from groundbreaking to occupancy, the structure took only 6 months and 11 days to complete.

One of the highlights on any architectural tour of the Loop is the 10-story 1889 Auditorium Building, designed by Louis Henri Sullivan and located on Michigan Avenue. A National Historic Landmark, the building originally included office space and a hotel, as well as a theater renowned for its intricate ornamentation and superb acoustics, the latter primarily the work of Sullivan's partner, Dankmar Adler. After years of neglect the building was given a $3-million face-lift in the 1960's. Roosevelt University now occupies the theater and maintains a small exhibit of Sullivan memorabilia in the old hotel lobby.

FOR MORE INFORMATION:

Chicago Architecture Foundation, 224 South Michigan Ave., Chicago, IL 60604-2501; 312-922-3432.

The Ravenswood El (short for elevated), left, circles the Loop's skyscrapers before heading north and west to Lawrence and Kimball streets.

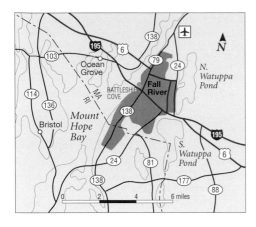

They called it Big Mamie and The Mighty Massy—and with good reason. Each of the nine 16-inch guns on the battleship *Massachusetts* can fire a shell weighing as much as a car for a distance of up to 20 miles. During almost five years of active service, the warship steamed 225,000 miles—the equivalent of traveling five times around the world. With a crew of more than 2,300 officers and enlisted men, the ship was a floating town. In 35 engagements with hostile forces, from the coast of North Africa to Tokyo Bay, not a single crew member lost his life in battle.

Today the guns of the battleship are trained fore and aft, signifying the vessel's peaceful intent. Moored since 1965 in Battleship Cove in Fall River, Massachusetts, the ship is the centerpiece of a museum of 20th-century naval history, as well as the official war memorial of the commonwealth of Massachusetts.

Surprisingly, the 43,000-ton, 680-foot long *Massachusetts* is small for a battleship when compared with ships built after it. But when launched, Big Mamie was the most formidable fighting platform in the world. As one of four South Dakota Class battleships, it was designed according to the size and weight limitations laid down by naval arms treaties signed following World War I. In order to keep the weight of armor down and yet still protect the ship, designers reduced the length of South Dakota Class ships and came up with innovative space-saving arrangements to house the engines, boilers, and other ship machinery.

Relative sizes notwithstanding, such battleships as the *Massachusetts* were weapons of staggering power, originally designed to pound an enemy fleet with a barrage of steel from their heavy armaments. There were few sights more awesome than the sight of a squadron of these gray-painted behemoths heaving through the waves. However, with the development of aircraft carriers in the 1930's, the role of the battleship changed to one of providing artillery cover for amphibious landings and muscle for fast carrier task forces.

On November 8, 1942, Big Mamie went into action in support of the Allied landings in North Africa. Soon after, in Casablanca Harbor, the ship's 16-inch guns were deployed to silence the Vichy-French battleship *Jean Bart*. From North Africa the *Massachusetts* traveled to the Pacific and saw action in the Coral Sea in carrier strikes against Palau, the Marianas, and the China coast. It also took part in the invasions of the Gilbert Islands, the Marshall Islands, New Guinea, and the Philippines. Some of the most furious combat experienced by the battleship took place during the invasions of Iwo Jima and Okinawa, when the *Massachusetts* had to fight off suicide attacks from Japanese kamikaze aircraft. Amazingly, the ship sustained its greatest damage not from the Japanese but from a typhoon that struck the U.S. Third Fleet off the Philippines on December 17, 1944.

In 1962 the *Massachusetts* was stricken from the Navy Register and ordered sold for scrap. But the devoted crew of Big Mamie mobilized to raise money to preserve the ship, and in June 1965 the *Massachusetts* was towed from Norfolk, Virginia, to its final mooring at Fall River. The arrival of the great battlewagon helped to revive Fall River's waterfront area and was the catalyst for the establishment of Fall River Heritage State Park.

ONE-OF-A-KIND MUSEUM

Battleship Cove has prospered since the arrival of the *Massachusetts,* and visitors can now explore a number of other naval vessels, as well as aircraft and related exhibits. In 1972 the ship was joined by the submarine U.S.S. *Lionfish;* the long-serving destroyer U.S.S. *Joseph P. Kennedy, Jr.*—named for the elder brother of Pres. John F. Kennedy—was placed on exhibit in 1974. To these were added PT 617 and PT 796, both fast wooden patrol boats once armed with torpedoes and used to attack enemy shipping. The collection also includes a Japanese suicide motorboat and a landing craft of the type used to put Allied troops and tanks ashore during World War II amphibious operations.

Among the aircraft on display are a T-28 Trainer plane and a Bell UH-1M Iroquois helicopter, better known as the Huey, which was used in the Vietnamese War. In 1996 the facility acquired the Soviet-built *Hiddensee,* a fast missile corvette originally designed to protect the coast of the former German Democratic Republic. Opened to the public in the spring of 1997, the sleek *Hiddensee* offers a fascinating glimpse of the Cold War from the other side.

FOR MORE INFORMATION:

Battleship Cove, Fall River, MA 02721; 508-678-1100 or 800-533-3194.

The Massachusetts, below, was the fourth vessel of the U.S. Navy to bear that name. It was built by the Bethlehem Steel Company at Quincy, Massachusetts, and launched on September 23, 1941.

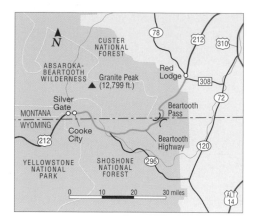

Travelers along the Beartooth Highway, left, pass numerous features christened by the workers who built the road. Some of the evocative names are Lunch Meadow, Mae West Curve, and High Lonesome Ridge.

The 68-mile-long Beartooth Highway snakes through a rugged terrain shaped by ancient volcanoes and slow-moving glaciers. Named for the Beartooth Mountains, this section of U.S. Highway 212—stretching between Red Lodge, Montana, and the northeast entrance of Yellowstone National Park at the town of Silver Gate, Montana—takes travelers on a three-hour drive through sweeping valleys and up over an 11,000-foot plateau as it passes towering peaks, alpine meadows, and glassy lakes.

Flanked by the 945,000-acre Absaroka-Beartooth Wilderness to the north and west, the Beartooth Highway was declared an engineering impossibility when it was first surveyed in the early years of this century. Although Native Americans had traveled in the area for hundreds of years, the route's first recorded travelers were Gen. Philip Sheridan and 129 of his soldiers, who marked out a route across the mountains from Cooke City to Billings in 1882. The route pioneered by Sheridan served as a trail for the local mining towns and later as a rough guideline for the modern Beartooth Highway.

EARLY PROMOTER

One of the highway's early promoters was Red Lodge physician J.C.F. Siegfriedt. Doctor Siegfriedt feared that someday the local coal mines would close and the economy would go into a slump. Seeking a way to bring tourists—and their money—into the area and through to Yellowstone, he lobbied for the construction of a road over the Beartooth Mountains.

After Siegfriedt won the necessary government support for the extension of the road between Billings and Cody, construction began in 1919. But the road would not be completed: because of its tortuous course, it was built only to the 13th switchback instead of to the top of the 11,000-foot Beartooth Plateau. The road soon fell into disrepair and was left abandoned.

Siegfriedt's fear for the future of Red Lodge materialized a few years later when the Red Lodge mines shut down. As a result, local businessmen sent *Red Lodge Picket* publisher O.H.P. Shelley to Washington, D.C., to lobby for funds to construct a road between Red Lodge and Cooke City. By the late 1920's Shelley had gained the support of Montana's congressional delegation, who sponsored the Park Approach Act. The act, which President Hoover signed into law in 1931, called for the construction of scenic routes to the country's national parks through federally owned land. A classic example of a pork barrel project, Beartooth Highway was the only road to be constructed under the act.

In 1931 work began on the Beartooth with $2.5 million Depression-era dollars earmarked for the project. It was completed in 1936. Because only an approximate route of the road had been worked out in a survey carried out seven years earlier, much of the construction was ad hoc. The work was brutal, and two men died in accidents.

The Beartooth Highway is generally kept open from Memorial Day to the middle of October, when the snow inhibits traffic. Drivers should gas up their cars before heading out—there are no gasoline stations between Red Lodge and Cooke City, and the only concession stand on the plateau is the Top of the World Store, about 35 miles from Red Lodge.

FOR MORE INFORMATION:
Red Lodge Area Chamber of Commerce, 601 North Broadway, Red Lodge, MT 59068; 406-446-1718.

The view of Red Lodge from East Bench, below, reveals a small town of about 2,300 residents that sits on the edge of Custer National Forest. Visitors can stay in a hotel or motel in town or camp in one of the campgrounds along the highway, including those at Island Lake and Beartooth Lake.

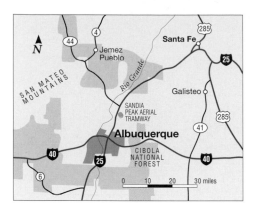

Few amusement park attractions can match the thrill of a ride on the Sandia Peak Aerial Tramway. In a mere 18 minutes the tramway whisks passengers some 4,000 feet from the terminal near Albuquerque up to the windswept summit of 10,378-foot Sandia Peak, site of a popular ski area in the Cibola National Forest.

The tramway is the longest in the world, measuring 2.7 miles long, with a clear span of 1.5 miles from the second tower to the terminal on the mountaintop. The wire rope tracks that carry the two cable cars are supported in only four places, a system that engineers call a double reversible jigback aerial tramway—jigback meaning that one car is pulled up as the other is being pulled down.

EUROPEAN MODEL

Robert Nordhaus, a member of the 10th Mountain Division during World War II and one of the founders and owners of the Sandia Peak Ski Company, came up with the idea for the tramway in 1961 while on a trip to Europe, where aerial tramways in ski areas are a common feature. For Sandia Peak, Nordhaus envisioned an aerial tramway that would shuttle skiers from the outskirts of Albuquerque to the ski slopes, thus eliminating the need for a half-hour drive along icy mountain roads.

Nordhaus hired Bell Engineering of Lucerne, Switzerland, to design the tramway and the Albuquerque firm of Martin & Luther as the general contractor to build it. Although the Swiss company had designed more than 50 aerial tramways in Europe, the firm declared that the Sandia Peak project was their most challenging.

The work began in May 1964 with construction workers building a half-mile service road up the mountain face to the 7,010-foot-high site that had been selected for the first tower. A massive crane was then hauled along the service road to erect the 232-foot tower, which was constructed to lean 18 degrees away from the hill to match the diagonal that would be taken by the lift line between the lower and the upper terminals. In order to reduce the size of the concrete footings from 350,000 pounds to 85,000 pounds, they anchored the tower to each concrete foundation with several 30-foot steel rods. This move produced substantial monetary savings at the time of construction: in the mid-1960's, the average cost of concrete was approximately $300 per cubic yard.

Although the second tower measured only 80 feet high, the rugged terrain and its elevation of 8,750 feet called for expensive and time-consuming work. Since no roads penetrated this area of the wilderness, all the workers and equipment had to be flown in by helicopter. Because a track-mounted air-drill used to drill the tower foundation anchor rods was too heavy to transport up the mountain in one piece, it had to be dismantled and the parts flown in by helicopter and later reassembled. Some of the tower sections were cut into still smaller pieces, then spliced at the site. It took 5,000 helicopter trips to complete the entire project. Despite these difficulties, the workers and engineers managed to keep the towers and terminals in proper alignment so that the line from one end to the other was off by only three-eighths of an inch.

Perhaps the greatest challenge tackled by the tramway's engineers was the stringing of the track cables between the terminals and towers. A team of 18 men had to run a three-eighths-inch cable down the mountainside and position it in a jigback arrangement to pull the heavier cables into place. Five months passed before the four track cables and the two hauling cables for pulling the coaches were secured.

Once the entire project was completed—a feat accomplished in only 24 months—it had cost its investors about $2 million; today, it would cost about 10 times that much. Before being opened to the public, the aerial tramway underwent 60 days of rigorous tests. After the tests proved successful, the first riders ascended Sandia Peak on May 7, 1966.

The Sandia Peak Aerial Tramway now shuttles an average of 275,000 passengers up the mountain each year. In the winter most of the visitors head straight for the ski hills. But in milder months, hikers explore Sandia Crest's 26 miles of trails, hang gliders launch themselves from its precipitous slopes, and diners enjoy the splendid view of pine forest and red canyon from a mountaintop restaurant.

FOR MORE INFORMATION:
Sandia Peak Aerial Tramway, 10 Tramway Loop NE, Albuquerque, NM 87122; 505-856-6419.

One of Sandia Peak's two cars, below, glides along its cables as the system is pulled by the 600-horsepower DC motor at the bottom. Each car can comfortably hold 55 people.

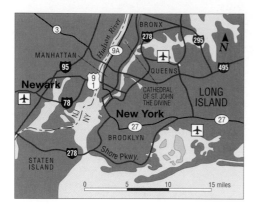

The Cathedral of St. John the Divine, above, was named after the author of the Bible's book of Revelations. The word "divine" refers not to the adjective, but to the noun, which means "theologian."

Workmen laid the cornerstone for the Cathedral of St. John the Divine on December 27, 1892. More than a century later the church in Manhattan's Morningside Heights remains unfinished. Yet even in its current state, it is the largest cathedral in the world. Among churches of all kinds, it ranks third behind a new basilica in the Ivory Coast and St. Peter's Basilica in Vatican City.

BIG JOHN

Fondly known as Big John to New Yorkers, the cathedral is the mother church of the Episcopal Diocese of New York and the seat of its bishop. An immense nave, measuring 248 feet long and 146 feet wide under a 124-foot-high ceiling, is located within the 610-foot-long interior. Three thousand people can sit comfortably in its seats in full view of the Tennessee marble pulpit, which has hosted a variety of speakers who range from the Dalai Lama and First Lady Hillary Clinton to actor Robert Redford, Nobel prize–winning African-American author Toni Morrison, and Sesame Street's Big Bird. Perhaps the interior's most striking feature is its eight granite columns—each the height of a five-story building and weighing 130 tons. The massive columns rise behind the main altar.

Church officials bought the 13-acre plot of land in Manhattan for $885,000 from the Leake and Watts Orphan Asylum. Their plan was to build a cathedral large enough to host the 1913 General Convention of the Episcopal Church. However, changing architectural taste, funding obstacles, and politics have slowed the construction from the beginning. The eastern portion of the cathedral was begun by Romanesque Revivalists from the firm Heins & LaFarge. In a fitting start to the cathedral's slow crawl to completion, workers dug for two years before they hit bedrock 70 feet below the surface. The western portion, opened in 1941, was conceived by Gothic Revivalists from the firm Cram & Ferguson.

In the 1970's, after a 30-year building hiatus, church officials tried to revive the cathedral building program, but discovered that the art of stone carving was almost extinct. In response, they set up a unique program to recruit master carvers from England, along with trainees from the cathedral's community. The apprentices, many of them local youth, have honed their craft carving stones for the western towers and central portal: along with stone carvings of Christian saints are Indian totem poles, New York skyscrapers, blue jays, bears, and likenesses of both Abraham Lincoln and Nelson Mandela. The completed statuary of the cathedral's central portal was dedicated in October 1997. During the years before the cathedral's carving program was ended in 1992, visitors to the cathedral were likely to hear the tap-tap-tap of carvers as they worked with the hand tools of a medieval cathedral builder: a hammer, punch, chisel, and riffler, or rasp.

Estimates of a completion date range from 50 years to never. Certainly millions of dollars would be needed to see the cathedral in its finished state. However, in the words of the dean of the cathedral, Harry H. Pritchett Jr.,"Although structurally incomplete, the Cathedral has become a world monument, known as much for its spiritual and intellectual light as for its soaring arches."

FOR MORE INFORMATION:
Cathedral of St. John the Divine, 1047 Amsterdam Ave., New York, NY, 10025; 212-316-7540.

The Great Rose Window over the west door of St. John the Divine, right, is made of more than 10,000 pieces of stained glass.

The Statue of Liberty, right, was built with 400,000 francs raised from the people of France by the Union Franco-Américaine.

She began as a gift from the people of France to commemorate America's centennial. But since her unveiling on October 28, 1886, the colossal statue in New York Harbor, *Liberty Enlightening the World*, has become a powerful symbol of freedom to people everywhere.

The idea for the statue was originally suggested in 1865 at a quiet dinner party in Versailles, France, by the party's host, teacher and jurist Edouard de Laboulaye. One of his guests was a young sculptor named Frédéric-Auguste Bartholdi, who seized on the idea and for the next 21 years worked obsessively to see it to fruition.

A COPPER COLOSSUS

Bartholdi had long admired 17th-century sculptor G. B. Crespi's statue of St. Carlo Borromeo at Lake Maggiore in Italy. This 76-foot-high statue had been built of beaten sheets of copper mounted on a framework of iron. Bartholdi decided to produce his towering statue in a similar way.

Work on the sculpture began in 1875 in the Paris factory of Gaget, Gauthier & Cie. A crew of French craftsmen determined the 151-foot Statue of Liberty's final dimensions by building three smaller clay models: the first clay figure measured just over 4 feet, the second, 9 feet, and the third, 36 feet. The workers took detailed measurements of everything from the tip of the nose to the broken shackles at Liberty's feet to provide the fourth and final version with the correct proportions.

These measurements were used to produce a full-scale wooden lath model of the statue, which was covered first by plaster and then by wooden molds. Craftsmen used a technique called repoussé, whereby they beat the copper into shape by hammering it into the wood molds. The result

was 300 copper sheets, a penny and a half thick, that perfectly conformed to all aspects of Lady Liberty's image, including the face, which Bartholdi had modeled after both his mother and his mistress.

Alexandre-Gustave Eiffel, a bold engineer and specialist in iron bridges, designed the iron framework for the sculpture using an almost 97-foot pylon made of four iron posts running upward from the statue's base. From the pylon he extended a series of secondary iron beams that conformed to the statue's shape. Still more iron beams supported the copper sheets that were further secured with flat, springy iron bars. Once the copper sheaths were in position, they were riveted together.

After the work was completed, the 225-ton statue was disassembled, crated, and shipped by steamer to New York City, where it spent 15 months in crates awaiting completion of the pedestal, which had encountered funding problems.

Designed by the American architect Richard Morris Hunt, the 89-foot-high concrete-and-granite pedestal was made with money donated by the American people, after Hungarian immigrant Joseph Pulitzer headed up a fund-raising campaign through his newspaper, *The World*. The

pedestal, which on its 65-foot base looks vaguely Egyptian, was made to enhance, rather than overpower, the statue.

Today regularly scheduled ferries to the statue and nearby Ellis Island leave from Manhattan and Jersey City, New Jersey. A museum in the base of the statue's pedestal traces Liberty's history. Hardy sightseers can climb the winding 354 steel steps from the bottom of the pedestal to the statue's crown, where they are rewarded with breathtaking views of the harbor and the skylines of Manhattan and Brooklyn through specially designed windows.

Over the years Liberty's torch was partially stripped of its copper sheathing and deformed by renovations. In 1984, as part of a refurbishment of the statue for the centennial, the torch was unbolted and lowered. French craftsmen traveled to New York and, in a fitting tribute to the work of their 19th-century compatriots, re-created Bartholdi's gleaming gold-leafed torch.

FOR MORE INFORMATION:
Statue of Liberty National Monument, Liberty Island, New York, NY 10004; 212-363-3200.

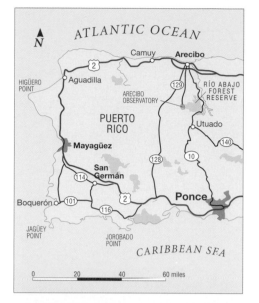

Steel cables support the more than 37,000 aluminum panels that make up the reflector dish, above. Beneath the panels, ferns, grasses, and wild orchids provide a habitat for lizards, snakes, and birds.

Hanging 450 feet above the static dish, the 75-ton Gregorian reflector system, above, tracks radio sources 10 billion light-years away.

Tucked neatly away in a natural sink-hole deep in the hills of northwestern Puerto Rico, a giant antenna gathers radio waves emitted billions of light-years ago by dying stars in the far reaches of the universe. The bowl-shaped Arecibo Observatory is the world's largest radio telescope, and the important data gathered from it by astronomers are helping to shape human beings' notions of the size and shape of the solar system.

Radio telescopes are the eyes and ears of radio astronomers. This branch of astronomy originated in the 1930's, when scientists at Bell Labs realized that radio waves originating from the center of the Milky Way were causing the static that was interfering with radio transmitters installed on ships at sea. In order to detect these energy emissions, Prof. William E. Gordon of Cornell University came up with the idea of building a giant antenna that would collect radio waves and reflect them to a movable receiver suspended overhead.

Because the sun, moon, and planets pass almost directly above the island of Puerto Rico, it was the perfect location for such a telescope. Completed in 1963, the Arecibo Observatory's reflector dish consists of 37,778 aluminum panels supported by steel cables running just above the floor of a sinkhole. The 1,000-foot-diameter reflector dish covers an area of 19 acres—roughly the equivalent of 17 football fields.

PROBING THE OUTER LIMITS
Administered by Cornell University in cooperation with the National Science Foundation, the Arecibo telescope is among the most sensitive instruments of its kind on earth. It has enabled scientists to discover objects such as pulsars (neutron stars formed during supernova explosions that emit strong pulsed radio waves) and quasars (the central regions of young galaxies). Arecibo is also a center for radar

astronomy, in which scientists beam signals at planets, moons, and other objects. These signals are then collected for study as they bounce back to earth. Radar images collected by Arecibo have revealed for the first time the cloud-shrouded surface of Venus. Much closer to home, the Arecibo radar is used to study the earth's ionosphere—the portion of the atmosphere more than 30 miles above the planet's surface.

Fascinated by this engineering wonder and by the research carried on here, thousands of amateur stargazers visit the observatory each year. Arecibo is a great source of pride for Puerto Ricans, and local contributions helped to fund the Fundacion Angel Ramos Visitor and Educational Facility—named after publisher and philanthropist Angel Ramos. The visitor center presents exhibits in English and Spanish on the exploration of the universe, and houses meeting rooms, a 100-seat multipurpose theater, and a science merchandise store. A viewing platform, located 200 feet above the reflector rim, provides a grandstand view of the telescope and surrounding hills. A cloud-flecked Caribbean sky hints at the limitless majesty of the universe beyond.

FOR MORE INFORMATION:
Arecibo Observatory, P.O. Box 995, Arecibo, PR 00613; 787-878-2612.

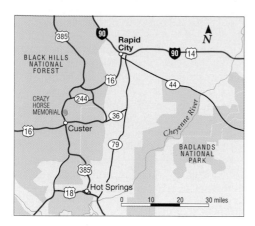

During the 1930's Native American leaders watched in silence as the faces of four presidents were carved into the rock of Mount Rushmore in the Black Hills, the ancestral homeland of the Lakota, formerly known as the Sioux. In 1939 one of these chiefs, Lakota Henry Standing Bear, wrote an appeal to Korczak Ziolkowski, a man who had worked with Gutzon Borglum on Mount Rushmore: "My fellow chiefs and I would like the white man to know that the red man has great heroes, too." These words made a profound impression on the self-taught sculptor of Polish descent.

Seven years later, Korczak (as Ziolkowski preferred to be called) accepted an invitation to create a monument of the Lakota leader Crazy Horse in honor of all North American Indian peoples. The sculpture was to be carved in the round from a mountain five miles north of Custer—a town named after George Armstrong Custer, the arch enemy of the Lakota who was killed by Crazy Horse's warriors at the Battle of the Little Bighorn in Montana.

Korczak's ambitious plan was to create the largest sculpture in the world—a 563-foot-high, 641-foot-long statue of Crazy Horse pointing over his stallion's head to the sacred Black Hills. The Lakota chief's head alone would be larger than all four of the 60-foot-high faces atop Mount Rushmore, situated only 17 miles away.

A BLAST SIGNALS THE START

The dedication blast took place on June 3, 1948, but it took two years of pioneering to build the roads and route electricity to the site before the project was up and running. During that time Korczak refused offers of federal grants, insisting that a Native American memorial should be financed privately and not by the government that had broken its treaties.

To help people visualize the monument, Korczak painted its huge outline on the mountainside, which took some 176 gallons of paint. Suspended from a rope 400 feet above the treetops, the sculptor used an army field telephone to communicate with his wife, Ruth, who helped fix his position from the studio one mile away.

To determine where to blast away the granite, measurements were taken from a scale model, then transferred to the mountain. The work was arduous, and only recently eased with the aid of computers.

Korczak spent the 1970's and early 1980's blocking out the 22-story horse's head, blasting away the entire right side of the mountain in the process. The difficult work of blasting and bulldozing took its toll, and over the years the sculptor underwent four back operations and a quadruple heart by-pass operation. On his deathbed in 1982, Korczak's parting words to his wife were these: "You must work on the mountain—but go slowly so you do it right."

Ruth and seven of their children took over the project, working from Korczak's scale models and three books of measurements that she had helped her husband prepare. In 1987 Ruth and the directors of the nonprofit Crazy Horse Memorial Foundation decided to shift the focus of the work efforts from the horse's head to Crazy Horse's relatively smaller head, predicting that public support would increase once detailed results were visible.

By 1997 some 8.4 million tons of granite had been blasted from the mountain. Today the sculpture-in-progress features the warrior's nearly completed face, the rough outline of an outstretched arm, and the rearing head of a stallion. The Ziolkowski family continues to work toward the goal of completing the face by June 3, 1998—the 50th anniversary of the memorial's dedication.

Korczak's goal was to preserve and promote Native American culture, and his vision encompassed much more than his massive sculpture. Today an ever-expanding collection of artifacts is housed in the temporary on-site Indian Museum of North America. A permanent museum, based on the circular design of a traditional Native American hogan, will be built at the base of the mountain when the Crazy Horse monument is completed.

Once the entire sculpture is finished—no deadline has been set yet—the equestrian figure will be more than twice the height of the Statue of Liberty. The mountainside will bear a poem by Korczak carved in stone, describing Crazy Horse as a proud warrior pointing to the Black Hills and proclaiming, "My lands are where my dead lie buried."

FOR MORE INFORMATION:

Crazy Horse Memorial, Ave. of the Chiefs, Crazy Horse, SD 57730-9506; 605-673-4681.

Crazy Horse's profile emerges from the rock, left. When completed, the opening beneath the figure's outstretched arm will be 10 stories high.

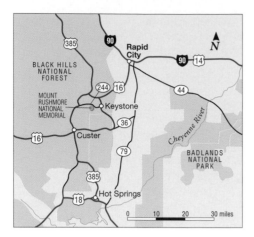

Four noble visages atop Mount Rushmore gaze out over the nation they once led, serving as tributes to the ideals upon which America is based: George Washington stands for independence, Thomas Jefferson for the democratic process, Abraham Lincoln for equality, and Theodore Roosevelt for the nation's leadership role in world affairs.

The colossal monument was the brainchild of state historian Doane Robinson, who in 1923 planned to draw sightseers to the region by having the statues carved in the Black Hills of South Dakota. Gutzon Borglum, an artist of Danish descent, accepted the Herculean task of designing and completing the project that has been dubbed the "shrine of democracy."

Born in Idaho in 1867, the sculptor studied art in San Francisco and in Paris, where he befriended French sculptor Auguste Rodin. Among Borglum's early works were a remodeled torch for the Statue of Liberty, saints and apostles for the Cathedral of St. John the Divine in New York City, a seated Abraham Lincoln in Newark, New Jersey, and a bust of Lincoln for the U.S. Capitol in Washington, D.C.

Borglum's patriotic vision and insistence that a big country demanded large-scale art helped transform Robinson's idea for a tourist attraction into an American icon. The artist proposed a sculpture of four of America's greatest statesmen and selected a sun-washed cliff atop 5,725-foot Mount Rushmore as his raw material. Borglum said, ". . . let us place there, carved high, as close to heaven as we can, the worlds of our leaders, their faces, to show posterity what manner of men they were."

Construction began in 1927 and lasted 14 years, though only less than half those years were spent on carving. The work was fraught with delays caused by poor weather and a lack of public support for the project, all of which was exacerbated by the economic conditions of the Great Depression. By personally lobbying congressmen, cabinet members, and presidents, Borglum eventually raised $836,000 in federal funds; the remaining $164,000 was acquired from various private sources.

DYNAMITE AND JACKHAMMERS
The only measuring system used for the gargantuan sculpture was Borglum's pointing machine, which consisted of a swivelling pointer on a protractor plate (a semicircular instrument used for measuring angles) and a weighted plumb line. Measurements were taken from five-foot models of each figure and then transferred to the face of the cliff in direct proportion, using a much larger pointer that was fastened to the mountaintop.

After locating a specific reference point on the face of the mountain, workers would then blast the granite to within inches of the finished surface. Seated in swing seats suspended by cables, workers used jackhammers to honeycomb the surface with shallow holes about three inches apart. They then dislodged the remaining rock with small drills and wedging tools. Finally, handy-sized, air-powered hammers were used to smooth the sculptures and enhance the reflective quality of the granite.

Each of the sculptures was given a lifelike appearance by adding finishing touches to their faces. These included Lincoln's beard, made by carving vertical lines into the rock, and Roosevelt's pince-nez, the faint outline of which encircles the great man's eyes. Borglum left a two-foot-long protruding shaft of granite in the eyes of each figure that reflects light and animates the pupils. The monument was all but completed when the sculptor died in March 1941 at the age of 74. His son, Lincoln, supervised the final touches to his father's masterpiece that same year.

The bits, chisels, and hammers used to carve the sculptures are preserved at Mount Rushmore National Memorial in Borglum's studio—the second studio he built for the project. An audiovisual presentation in the visitor center details the history of the project, and a viewing terrace offers views of the colossal faces hewn in stone.

FOR MORE INFORMATION:
Mount Rushmore National Memorial, P.O. Box 268, Keystone, SD 57751; 605-574-2523.

In all, some 450,000 tons of rock were excavated to create the Mount Rushmore sculpture, below.

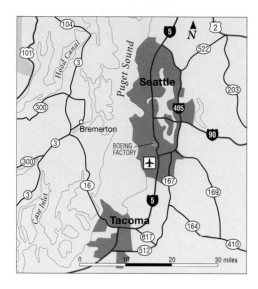

A phalanx of brand new 747-400's, above, awaits testing at the company's Everett plant.

Seven years after the Wright brothers' historic first flight, *Scientific American* magazine stated, "To affirm that the airplane is going to revolutionize the future is to be guilty of the wildest exaggeration." At the time, aircraft were rickety and unreliable devices barely capable of lifting their own weight. People saw them more as a source of amusement than a viable form of transportation. But a 28-year-old Seattle timberman named William Boeing had different ideas. Boeing believed sturdier and more dependable aircraft would forever change the face of transportation. Flying on little more than a wish and a prayer, Boeing and a friend founded Pacific Aero Products Company in Seattle on July 15, 1916.

Boeing's fledgling business literally took off during World War I. Navy orders for Boeing's planes helped the company grow until the 1918 armistice, when the military cut back on production. After the war, Boeing decided to focus his business on airmail deliveries. The decision almost led to bankruptcy for his company, and Boeing was forced to personally come up with the finances needed to keep the company alive. At one point, he even turned to building boats and furniture just to keep the doors of his business open.

As often is the case, war brought prosperity to manufacturers, and Boeing was no exception. During World War II the company manufactured more than 13,000 B-17 bombers. The Boeing plant also produced B-29 bombers, which were active in the Pacific Theater during the war.

Following the war the company continued to build military aircraft, but it also directed much of its energy toward commercial transportation. Long on the cutting edge of technology, Boeing's engineers helped lead the change from propeller-driven aircraft to jet airplanes; it produced its first jet-powered passenger plane, the 707, in 1958. Today commercial planes make up 70 percent of Boeing's business.

BIG FACILITY FOR BIG PLANES

In the late 1960's Boeing established itself as the world leader in aircraft manufacturing with the opening of its cavernous 747 plant in Everett, located just 30 miles north of Seattle. Making the largest commercial jetliner naturally requires a large building, and at 472 million cubic feet, the Everett plant is easily the largest building by volume in the world. The sprawling facility covers 98.36 acres of land and employs 14,000 people—some of whom need bicycles to reach different parts of the building. The electricity used to light the plant could supply light to 32,000 average American homes. It serves as a fitting birthplace to some 100 Boeing 747's each year.

A tour of the enormous plant gives visitors a look at the intricate process that goes into the construction of a 747. A short film condenses the nearly two months it takes to build one of these mammoth aircraft into seven minutes. After viewing the film, guides lead visitors to the manufacturing plant through an underground corridor. A short elevator ride takes sightseers to an observation balcony that overlooks the gleaming noses, center sections, and tails that are awaiting assembly. Nearby, visitors can watch workers rivet together some of the 4.5 million parts that make up each aircraft. Not far from this, a completed plane awaiting its first flight stands by a door as wide as a football field.

During the tour visitors learn a number of interesting facts about the 747 that illustrate both the care that goes into making it and the enormity of the finished product. The first B-747 required 75,000 engineering drawings before completion. That plane was entered into service on January 21, 1970, and has since carried 1.8 billion passengers more than 24.7 billion miles. The tail height alone of a 747 is equal to a six-story building. How far has aircraft design come since the industry's earliest days? The Wright brothers' first flight could have been performed within the 150-foot economy section of the B-747-400—Boeing's newest and largest plane.

Today the Boeing Company employs more than 227,000 people—the state's largest private employer—and records $22.7 billion in sales. It delivers 220 airplanes a year, nearly 100 of which come from its Everett plant. Despite the early words of foreboding, Boeing's dream proved well within the bounds of reason.

FOR MORE INFORMATION:

Boeing Tour Center, Boeing Commercial Airplanes, Seattle, WA 98124; 206-544-1264.

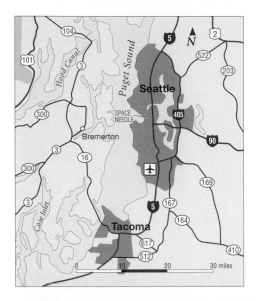

On April 17, 1961—just one year before the opening of the fair—workers began to excavate a hole for the foundation, which measured 30 feet deep and 120 feet across. The base of the foundation was reinforced with 250 tons of steel bars, and 72 massive anchor bolts secured the steel legs and central tower firmly to the ground. In all, 467 truckloads of cement totaling 5,850 tons were used in the foundation.

Weighing as much as the needle itself, the huge foundation was designed to lower the tower's center of gravity to just above ground level, thus increasing its resistance to lightning, earthquakes, and strong winds.

The Space Needle, below, has been painted three times since its completion in 1962—a job that requires some 1,000 gallons of paint.

Sections of the central tower were designed to be easily removed so that the 42-ton crane could inch its way back down again.

Both of the tower legs were welded together with massive spans of steel, each piece measuring 90 feet long and weighing 27,000 pounds. The three outer legs split at the 373-foot level to provide six outer supporting points for the restaurant and observation deck at the 520-foot level.

The top house and dome were so perfectly balanced that the track-mounted restaurant rotates with only a one-horsepower electric motor—the power needed to run a household vacuum cleaner. Hoisted and installed in 40,000-pound pieces, the giant turntable that supports the restaurant was engineered to make one complete revolution each hour. This 14-foot-wide ring of steel serves as the base of the top house and includes decorative rays with sun louvers that limit the amount of sunlight shining through the windows. One of the most dangerous jobs was bolting these sun louvers into place: with nothing to stand on or hold on to, workers had to inch their way out along the six-inch-wide, downward sloping steel rays, reaching out 515 feet above street level to attach the 75-pound pieces into place.

Built for only $4.5 million, the structure was completed in December 1961 and officially opened on the first day of the World's Fair—April 21, 1962. Ironically, the Space Needle's first manager, Hoge Sullivan, suffered from acrophobia—fear of heights.

More than three decades since its opening, the Space Needle is still a top tourist attraction, with some 1 million people a year making the ascent to the observation deck 520 feet above the ground. From this lofty perch they are afforded a spectacular 360-degree panorama of the surrounding area, a view that takes in the Olympic and Cascade mountain ranges, Mount Rainier, the port of Seattle waterfront, Puget Sound, and Lake Union. The outdoor deck is equipped with high-powered telescopes for more detailed sightseeing.

Milestones in the Space Needle's colorful past include a 1974 birth in a women's restroom and a pair of unauthorized parachute jumps. The scene of numerous weddings, conventions, and birthday celebrations, the middle-aged "spacecraft" is now firmly grounded in Seattle's history.

FOR MORE INFORMATION:
Space Needle, 219 Fourth Ave. N, Seattle, WA 98109; 206-443-2111 or 800-937-9582.

Poking above the Seattle skyline, the 605-foot-high Space Needle resembles a flying saucer balanced on a slender tripod. The tower was designed as a space-age landmark for the 1962 World's Fair, the theme of which was Century 21.

The original design, reputedly inspired by the Stuttgart Tower in Germany, was drawn on a paper napkin in 1959 by Edward E. Carlson, then president of Western International Hotels and a fundraiser for the World's Fair. Enthused by Carlson's plan, organizers of the fair hired Seattle architect John Graham, designer of a revolving restaurant in Honolulu, Hawaii, to spearhead the project.

The Space Needle was equipped with 25 lightning rods and engineered to withstand a wind velocity of 200 miles an hour—doubling the 1962 building code requirements.

Aboveground construction began with the central core tower, including two sets of stairs that climb 832 steps from the basement to the restaurant, and all the pipes, wires, and cables for the tower's mechanical systems and three elevators.

GROWING PAINS
Since the building would eventually outgrow the tallest crane, a special crane was designed to fit inside the central tower and lift itself up as the tower was built around it.

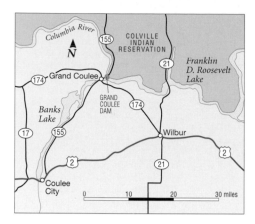

The Grand Coulee Dam generates enough hydroelectric power to serve more than 2 million households annually. Made of 12 million cubic yards of concrete, the gigantic dam is the country's largest concrete structure and king of the Columbia River Power Complex, a series of dams that tames the waters of the nation's second-largest river.

Located near Grand Coulee, Washington, the massive powerhouse was the driving force behind the economic boom of the Pacific Northwest following World War II. The building of the dam spawned such industries as aluminum smelters and wood processing plants; when completed, it irrigated farmland and orchards.

The gargantuan project was undertaken by Franklin D. Roosevelt's Depression-era Public Works Administration, a program designed to provide jobs for the unemployed through the development of the nation's resources. Preliminary engineering work began in 1933, and in 1934 Roosevelt visited the dam site to oversee the ceremonies marking the start of construction.

Two U-shaped cofferdams (temporary dams) were built on either side of the river to narrow the course of the Columbia and expose the bedrock below. After the water was pumped from the areas inside the cofferdams, massive amounts of sand, gravel, and boulders were funneled out of the channel via a conveyor-belt system.

The exposed granitic rock served as a foundation for the network of steel trestles that were built to support the body of the dam. Concrete columns grew five feet at a time; the crew waited 72 hours between each pour to allow the concrete to cure.

The dam was constructed outward from each bank toward the middle, with unfin-

The 660 miles of shoreline around Franklin D. Roosevelt Lake, right, are dotted with some 35 campgrounds and several hiking trails.

ished sections left as spillways. Two more cofferdams were built above and below the dam to divert the river over the spillways, allowing workers to prepare the central part of the riverbed for construction.

Construction ended in 1942. The dam measured 4,173 feet long and was later expanded to 5,223 feet—just 57 feet short of a mile. Enough water was harnessed by the 550-foot-high dam to create a reservoir, named Franklin D. Roosevelt Lake, which stretches 150 miles to the Canadian border.

Water is pumped from the lake and diverted south into a secondary reservoir at

Power lines cross the upper dam, left. Grand Coulee is the largest producer of hydroelectricity in the nation and the third largest in the world.

Grand Coulee, an immense natural channel formed when the Columbia was temporarily rerouted by glaciers during the last ice age. This secondary reservoir is drained by canals that supply irrigation water for more than half a million acres of farmland.

LASERS ON A LIQUID SCREEN

Besides quenching the thirst of area soils, the Grand Coulee Dam provided the impetus for the construction of numerous aircraft manufacturing facilities in the region during World War II. It was also critical to the activities at the Hanford Reservation, where much of the work on the first atomic bombs took place.

Visitors can watch the gigantic spillway from the Visitor Arrival Center, which overlooks the dam complex. The complex consists of the pump generator plant, three power plants, and the dam itself. On summer nights the cascading waters serve as a huge projection screen for a laser light show originating from a booth inside the center. The 36-minute-long program features brilliant images of galloping horses, leaping fish, and soaring eagles, accompanied by a narrated history of the Grand Coulee Dam.

FOR MORE INFORMATION:
Grand Coulee Project Office, Code 1400, P.O. Box 620, Grand Coulee, WA 99133; 509-633-9265.

INDEX

PICTURE CREDITS

ACKNOWLEDGMENTS

Cartography: DPR Inc.; map resource base courtesy of the USGS; shaded relief courtesy of the USGS and the National Park Service.

The editors would also like to thank the following: Lorraine Doré, Pascale Hueber, and Valery Pigeon-Dumas.